RECYCLING OUR FUTURE
A Global Strategy

RANJIT S. BAXI BSc, MBA, FRSA

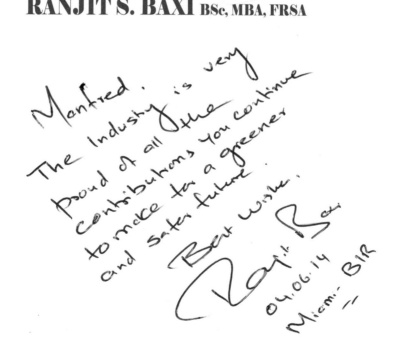

Monfred.
The Industry is very
proud of all the
contributions you continue
to make to a greener
and safer future.
Best wishes.
Rajit Ban
04.06.14
Miami - BIR

Whittles Publishing

Published by
Whittles Publishing,
Dunbeath,
Caithness KW6 6EG,
Scotland, UK

www.whittlespublishing.com

© 2014 Ranjit S Baxi

ISBN 978-184995-138-8

Printed and bound in DS Smith

Contents

Acknowledgements

Although this book is my opinion about the waste industry, its operation, direction and future, I must acknowledge the support I have had for the project from so many including my colleagues at the Bureau of International Recycling (BIR) of which I have been a proud member for more than 15 years. I should also pay tribute to the recycling organisations around the world including the Institute of Scrap Recycling Industries, Inc. (ISRI) the 'Voice of Recycling' in America that represents more than 1,600 companies.

I would like to acknowledge the support I have had over the last three decades from all my customers, contacts and friends in Europe, America and Asia.

In particular I would like to thank Björn Grufman, BIR President, Dominique Maguin, former BIR President, Alexandre Delacoux, BIR Director General, Reinhold Schmidt, President of the BIR Paper Division, Robert Stein, President of the BIR Non Ferrous Division, Christian Rubach, President of the BIR Ferrous Division, Olaf Rintsch, President of the BIR Textile Division, Robert Voss, Chairman of BIR's International Trade Council, and Francis Veys, former Director General of BIR for their contributions. And I would also like to thank many others who are unnamed for their encouragement as I have worked on the book.

I have quoted liberally from various publications and I trust I have given due credit on every occasion. I do know however that we are all driving towards the same goal: a greater understanding and acceptance of all methods of recycling to save our rapidly depleting finite resources.

I dedicate this book to my late mother and father who supported and inspired me as I set out on my career. Above all I must thank my wife, Harvinder, and my children Jasmeet, Harmeet and Aneet for the strength and support that they have always given me as I built my company J & H Sales (International) Ltd.

Lastly my thanks go to my publisher, Keith Whittles and his team at Whittles Publishing for their patience, guidance and skill.

Introduction

Dust to dust, ashes to ashes. There is a certain irony that man, the world's greatest polluter, has a most efficient in-built design which helps the body begin decomposing within days of being lowered into the grave and yet what he has created takes sometimes hundreds of years to disintegrate and even then it can 'live' on. Marine experts, for example, are convinced that we humans are consuming plastic when we eat fish because although all the plastic detritus that finds its way into the oceans eventually breaks down, it ends up as miniscule plastic pieces which release toxic chemicals (Bisphenol A and PS Oligomer) that the fish eat. Until we are able to build a natural obsolescence into everything we make with the same efficiency as our own self-destruct mechanism then there will always be waste.

Everything we do just to get through a normal working day contributes to the tonnes of waste material that has to be dealt with one way or another. According to the United Nations we produce more than one billion tonnes of domestic waste worldwide per annum which excludes manufacturing, construction, water supply and energy waste. In Europe alone if we include those waste streams the figure tops three billion tonnes.[1]

There is another phrase which might easily be applied to waste: out of sight, out of mind. Most of us believe that we are doing our bit 'recycling' our household waste into the multi-coloured bins which local authorities provide; we may even feel a sense of satisfaction which somehow justifies our filling the bins again the following week; and once that rubbish or garbage has been collected we do not want to know any more about it. Of course we are not recycling at all, we are merely pickers, the first link in a long chain of sorting and sifting before the real work of recycling begins and none of that must take place anywhere near our homes in case it offends our sense of smell. Then there is also the scandal about how much of our carefully sorted waste simply gets shipped off to far-flung destinations not to be recycled but to be dumped on poisonous, mountainous landfills where the poor and the desperate pick over them to scavenge a livelihood.

---

1   European Commission – Being wise with waste: the EU's approach to waste management

This book I hope will provide an insight for the layman into the workings of the waste industry which can be highly sophisticated despite its public image, and why it matters to every one of us that we know what is happening and what is at stake; to start thinking of the industry not as a disposer of waste but as a provider of resources.

We will be addressing common threads throughout the industry without dealing in fine detail with the intricacies of the various sectors – how to extract gold and platinum from plastic computer motherboards or the inner workings of an anaerobic digester. There may be unique problems related to certain sectors but looking across the globe, recycling can be divided into various broad groups: paper, plastic, glass, ferrous and non-ferrous, steel, textiles, tyres, plastics, electronics. These are the main branches of recycling, but this book is about the forces and drivers impacting the recycling business and what we must do together to overcome those hurdles rather than dwelling too much on the technology. There is nothing new about the need to recycle as my colleague, Dominique Maguin, former President of the Bureau of International Recycling (BIR)[2] reminds us: 'Man has always reused materials…in the Bronze Age, men were recovering broken items to redesign and make new ones.'

I also present some challenges for the industry and the legislators who in their desire to develop a new technology or the pursuit of tighter regulatory controls lose sight of what we are trying to achieve. The uncompromising drive for zero waste or local authorities striving for that extra buck will not in themselves result in the protection of our dwindling finite resources or a reduction in toxic waste mountains nor will they help our industry.

If we understand exactly how the recycling system works, who is benefiting and who is suffering as a result of our consumption, then we will at least be better informed and perhaps take different decisions not only about what we consume but also how we deal with what we discard. In particular I would like to discuss the challenges of the present day rather than dwell on the history of how we came to the crisis we now face.

This is not a lecture about human wastefulness – that is for other people – but it is about practicalities; as long as we continue to behave as we have done for centuries we will have to do something about what we throw away. It doesn't just disappear; it doesn't all naturally rot away, so we will definitely have to devise new ways of tackling the problem which we have managed to do for centuries, from the first scavengers picking over waste dumped in the streets to today's $200 billion plus per annum recycling industry employing some 1.6

---

2   The Bureau of International Recycling is a non-profit organisation founded in 1948 under Belgian law. It was the first federation to support the interests of the recycling industry on an international scale and represents companies and national associations in more than 70 countries.

million people.[3] In fact some calculate that the value of the industry, including municipal, industrial recycling and waste to energy, is many times greater, perhaps as much as $1 trillion a year with the potential to double by 2020.[4]

Many will be surprised that there is not even agreement about the best way to collect waste from our homes; we do it differently within a small country like the United Kingdom, we do it differently again in Europe and in America and elsewhere. The headlines are reserved for the next best thing – perhaps an innovative waste-to-energy technique – the investment money piles in and then it, too, is discarded for another system. But I would argue that our primary focus should be much earlier in the chain.

I have spent more than 30 years buying waste from companies that sort and pack it into containers and sell it on to people who do the actual recycling. The demands of the ultimate buyers, the mill owners for example who turn our waste paper and cardboard into usable product, change constantly and by that I mean daily; prices rise and fall as swiftly as the stock markets and the waste streams are many and varied, from whole ships and demolished buildings to a single battery or plastic bag. It may also come as a surprise to some that the very best quality waste is in short supply despite the waste mountains we all complain about and the apparent shortage of landfill sites. So curiously some waste is itself becoming a scarce commodity just like the finite resources we greedily consume and as such has a rising value.

What shocks me after so many years in this industry is how little people (who should know better) understand about waste: investors are hypnotised by technology without appreciating the basic requirement – quality; many in the industry seem to have little knowledge or firsthand expertise about the markets into which they are selling; governments and local authorities have for years failed to grasp the value of the one raw material which will never run out and seem willing to give away the potential profit by outsourcing the problem to contractors who can't believe their luck; and why do householders seem to be so willing to act as unpaid sorters on domestic picking lines in their own homes? When you screw up a ball of paper and throw it into your waste paper basket, pause for a moment and consider not only where that sheet of paper came from but about the journey it may be about to embark on. If you are careful about how you separate your waste paper it may soon be on the high seas heading for China, which accounted for 25 per cent of the world demand and 26 per cent of global production for all paper and board in 2012 retaining its position as the world leader for the fourth consecutive year.[5] And yet, times are changing.

---

3   Bureau of International Recycling (BIR)
4   Bank of America Merrill Lynch 2013 report: *No time to waste – global waste primer*
5   *2013 Annual Review of Global Pulp & Paper Statistics* – Resource Information Systems Inc

The traditional dumping grounds of our waste, usually Eastern and African destinations, are saying no more as they too begin to favour quality over quantity. This is having an impact on the whole industry and will in time no doubt be felt at householder level as we adapt to new conditions in the global market.

There is much that needs to change as we consume our natural resources faster than they can be replaced as the world's population grows, becomes increasingly affluent and therefore ever more demanding and wasteful. Developing nations want to enjoy the high life just as much as the rest of us have done for years and who has a right to stop them? We can shout about the need for a zero waste policy but we must recognise and allow for growth. Legislation, one of the most influential drivers of what we do with our waste, is in disarray; someone described waste management generally to be in a state of anarchy. It is already late but we must start Recycling Our Future and we can only do that by understanding how the process works.

1

Overview

It is an easy enough challenge: how are we going to clear up the waste we create?

To err is human, so the short answer is that we can't or won't reduce our waste and there are a number of reasons for that beyond our human frailties. The world's population is growing, albeit at a slightly reduced rate, and perhaps more importantly the middle classes are expanding; people in the fastest growing countries in Brazil, the Near, Middle and Far East, and China as well as parts of Africa are becoming increasingly affluent so they too want to share in the good times; between 1980 and 2009 the world's average GDP per capita grew by 248 per cent from $2,472 to $8,599 and after a sustained period of growth there is reported to be a big increase in African billionaires predominantly in South Africa and Nigeria.[6] Even though the International Monetary Fund reported a cooling in the growing economies of the so-called BRIC countries (Brazil, Russia, India, China) they still predicted growth of 4.5 per cent in 2013/14.[7] So if we won't consume less what should we do with what we are throwing away?

Before we look at techniques there is another issue which is arguably more important: our natural resources are said to be declining faster than they can be replenished and the most obvious example is our trees. In the so-called electronic paperless world we are actually using more paper than ever before and without recycled product we would soon not have enough virgin stock to meet demand. By far the greatest volume in any landfill is paper. Everyone worries about the plastic packaging which wraps everything from a single apple to a whole refrigerator, but plastic does not take up much room in a typical landfill. Dig deep enough and there will be old newspapers and telephone directories which have not rotted away as we fondly believe but are actually perfectly readable sometimes years after they were dumped.

---

6   *Venture Magazine* listed 55 billionaires in Africa up from 16 in previous estimates, *Financial Times* 7 October 2013

7   *Daily Telegraph* 9 October 2013

Into this mix we can throw the conflicting priorities of all involved in the recycling process which includes everyone from the householder to the end user or manufacturers of recycled products and their customers, with political bodies overseeing everything we do. As householders by and large we just want our dustbins emptied; it would be nice to know that what we have painstakingly sorted eventually gets recycled as opposed to being land filled but the priority is definitely that regular weekly collection.

The cities, councils and local authorities around the world responsible for collecting the waste, under constant pressure to save money, are inclined to go for the highest tender offer when they are looking for a contractor who will operate the service. Their priority is to have the waste collected efficiently, regularly and cleanly; what happens to the waste once the bins are emptied is largely of secondary importance so long as they are abiding by all the rules handed down from government. An alternative, less well known priority may be to make money. A spokesman for the UK's Department for Environment, Food and Rural Affairs (DEFRA) when asked about the vast increase in waste being sent abroad said:

> Trade in recyclable materials is a global market and we want to see UK businesses make money from it to help boost our economy. We would like to see our own recycling industry grow so that we can grasp this opportunity with both hands.[8]

As a little aside, it is worth remembering that it is the householder who is providing much of the 'raw material' for this money-making opportunity.

However the contractors will have their own preference as to how the waste is collected depending on the technology in their waste recycling facilities – some will want a high degree of pre-sorting by the householder while others such as in some cities in America will be quite happy for all waste to be dumped in one 96-gallon bin for separation later.

The argument over single-stream collection continues. It is not good enough to say collection rates have increased, which is itself debatable in the long run, therefore single streaming is the answer. Unless the Materials Recycling Facilities (MRF) are capable of improving the quality of their output it will remain no more than another pile of junk, albeit one which has been partially sorted and prepared.

To have a single-stream system can only work if the facility is capable of producing high quality recyclates which probably means more investment in equipment at the MRF and more staffing, and I would argue that in many

---

8   *Daily Mail* 4 September 2013

countries it should also mean more MRFs as well. The cost benefits are not clear cut. In some cases, in remote locations where it is impractical and uneconomic to have multiple collections, single stream may be the only sensible answer.

It is worth pausing for a moment to follow how we reached this present conundrum – single or dual-stream collection. In the late 20th century it began to dawn on people that we could not go on dumping our waste in landfills no matter how much land we thought we had; in Saudi Arabia for example they assumed they had plenty of desert to spare and drove out of town to dump their waste in the sand. Of course new prosperity has meant cities have grown and suddenly they found that they were running up against the landfills which now had to be 'mined' to make way for new buildings, so people began to recognise that the problem could not simply be buried.

In time as we became concerned about the ozone layer and greenhouse gas emissions, landfills became one of the targets and there was increasing talk of waste-to-energy. People had to be encouraged to recycle more so single-stream collection was pushed, but it doesn't take much to work out that by mixing up everything from food waste and paper to glass and cans results in highly contaminated waste; this is supposed to be the raw material from which new products are made in the actual recycling process. This was also a time of rapid economic growth which saw a huge demand in particular for pulp and recycled paper in the Far East – quality was not the key issue but that attitude would change. In addition the tempting economies of single-stream collection did not necessarily stack up. A study in 2002 by Eureka Recycling in Minnesota into different collection methods revealed that single-stream collection resulted in 21 per cent more waste being collected but the lower collection costs were 'outweighed by higher processing costs and lower material revenues'.[9]

But clearly not everyone is convinced. In 2013 in Montgomery, Alabama residents were told that they would soon be able to place all their waste in one bin which would be shipped off to a state of the art 'dirty' MRF which was being built at a cost of $35 million. It was forecast that the city's recycling rate would 'skyrocket' from its lowly 14 per cent. They were not alone. That same year the citizens of Minneapolis received their 96-gallon one-sort carts which could hold 200 lbs of waste that would be collected every other week. Others have been following suit.

However paper which is left outside overnight, even if it doesn't rain, loses its value because of the moisture content in the air that it soaks up. If it is left for two weeks at the bottom of a 96-gallon cart covered in liquids and every other type of waste it will have a zero value to end users. And no matter how

9   Container Recycling Institute – Understanding economic and environmental impacts of single-stream collection systems. December 2009

carefully glass bottles are placed in a waste bin, when it is automatically tipped into a cart, crushed and then dumped at the MRF for recycling it will render the paper worthless for recycling because of glass and liquid contamination. But the counter argument is that if single-stream facilities can improve the technology to process the input at a lower cost, then this might be an answer.

An almost diametrically opposite approach is being taken by some councils in the UK who have begun introducing smaller 140 litre wheelie bins with a fortnightly collection, arguing that the tighter restrictions will ultimately lead to greater recycling habits by households.[10] It is an interesting approach which has provoked protest from those who say the new regime will simply lead to more fly-tipping. The counter argument is instead of having a 'right' to weekly collections there should be a responsibility to recycle and reuse more.

I referred to dirty MRFs. This sounds like a pejorative term but it is simply shorthand for mixed unsorted waste. I will look at the recycling process in more detail later but suffice it to say here that mixed, multi-stream waste recycling facilities with the benefit of all the latest technology are certainly part of the answer, but the reality is that there are not enough of them and simply bolting on a new piece of kit to turn a modest facility into a 'super sorter' will not solve our problems.

The Container Recycling Institute offers some useful statistics: in the USA there are more than 160 single-stream MRFs and 27 per cent of the US population participate in single-stream programmes. This worked well to meet the demand of the explosive growth in the Far East when volume and not quality of material was a paramount issue, but now they are getting picky. It costs $5 to $13 more per tonne to process material from single-stream systems than from more sophisticated and cleaner facilities so obviously that is a cost they want to avoid. Some 40 per cent of glass from single-stream processes ends up on landfill – something we have long wanted to halt – while 90 per cent of glass from dual-stream systems is recycled.

As for the optimistic hopes that waste recycling will 'skyrocket' in cities with 96-gallon carts, the facts do not seem to bear this out. An analysis of three single-stream and four dual-stream systems showed that the weighted average costs of recovery were virtually identical for both – 6.9 per cent and 7 per cent – and the cost savings were as little as $0–3 per tonne.[11]

Collection is not recycling. Waste is only recycled when it can be returned to its original purpose or used in some other way and not end up perhaps as landfill cover, which is termed 'downcycling'. The aim is to achieve what is called

10   'One in four local authorities is now supplying bins that are up to 50 per cent smaller than those previ-
     ously given to households', *Daily Telegraph* 15 January 2014

11   Daniel Lantz, Metro Waste Paper Recovery, Ontario 2008

closed-loop recycling where items like a glass bottle can be cleaned and remade into a glass container or fibre glass indefinitely – it will never see the landfill.

So we have to be practical and assess the costs and benefits of everything from new monster bins to state-of-the-art facilities. The rate of recycling has to increase across the board and it will certainly involve a combination of single-stream and dual-stream collections. Technology is advancing which will improve the quality of product emerging from single-stream collection methods but as the chart of one US study shows (see Figure 1), a big difference remains, with single-stream collection still resulting in a high number of rejects.

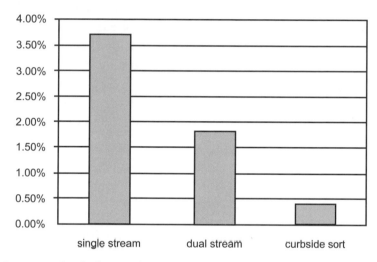

Figure 1 Percentage level of rejects from incoming material from various collections methods. Source: Targeted statewide waste characterization study (R.W. Beck, 2006)

There are other links in the chain which we will discuss but ultimately the waste ideally ends up at a reprocessing facility where the actual recycling takes place and the recovered fibre for example may be turned into boxes. The priority here is quality of recyclates and it is this priority which should be paramount and which is in fact increasingly determining how we manage our collections. Unless recyclables are free from contamination they will and are being rejected by the reprocessing companies who are looking to markets where collection quality is key; Europe is not at the top of this particular tree. In 2012 China, which imports 70 per cent of the world's electronic waste and 12 million tonnes of plastic waste each year, imposed its Operation Green Fence campaign that set a limit of 1.5 per cent of allowable contaminant for each bale of imported recyclables and it wasn't long before stringent controls were being applied to ensure importing and exporting companies were meeting the required standards.

Waste is an ongoing problem as well as being an asset from which some can profit. It is generated continually whether we are living in towns and cities or in the country; around the world everyone is generating waste. Historically the problem was dealt with simply by dumping what we didn't want in landfills which were no more than holes in the ground and before that of course we just threw it out into the streets. Waste has always come from multiple sources: commercial industry, households and hospitals and from industries producing toxic and hazardous waste. Furthermore all that waste was mixed together completely unsorted, leaching into the ground and surrounding soil. In the last 30 or 40 years the world realised that we have a finite source of the raw materials that are used to make products so we had to consider recycling or rather a more sophisticated method of recycling than picking over the dumps, although this continues today in some parts of the world. Since the 1970s we have increased the amount we recycle but we are not keeping up nor do we always focus on the most important priority: the quality of the recyclates we are sending to the end user. But recycling is helping in reducing the use of natural resources and consumption of energy compared with the production processes using virgin material (see Table 1)

Table 1

Energy Savings		CO_2 Savings	
Aluminium	> 95%	Aluminium	>92%
Copper	> 85%	Copper	>65%
Plastic	> 80%	Plastic	>58%
Paper	> 65%	Paper	>18%
Steel	> 74%	Steel	>90%
Zinc	> 60%	Zinc	>76%
Lead	> 65%	Lead	>99%
		Tin	>99%

One of the challenges today is for the waste generator to understand how the end user of that waste can best utilise it with the minimum cost on all sides. The aim of course is for 100 per cent usage of what we send to a reprocessor, but if we don't supply it in the right format largely free from contaminants then the waste will either have to be retreated or thrown in a landfill. Every country in the world has to recognise the importance of quality and make it a universal priority. Only by understanding where the waste is going, what specification is required and by adjusting our ability to deliver that level of quality can we truly make recycling work efficiently.

We also need to minimise logistics, in other words once the waste is collected it should ideally go to one point where the entire treatment of that waste takes place. Too often waste is taken to what is known as a transfer station where it may only be sorted, not even treated, and from there it is sent to further processing plants. This defeats the object of the exercise, it increases costs through transport, wastes energy in fuel and adds to the pollution of the planet. This is not recycling as it should be conducted but having said that big efforts are being made – the question is how effective are they.

We have endeavoured in the last 10 or 20 years to do more. Initially the challenge was the apparent scarcity of the finite raw material; however, despite apocalyptic warnings oil is still plentiful because we are extracting and using it more efficiently and we are even growing more trees than we are cutting down by a factor of 20[12] – it is just that we are felling trees which should be protected such as in the Amazon Rainforest.

Post-1995 and especially post-2000 the price of non-ferrous metal – copper, aluminium, platinum – increased dramatically so it became more important for those industries to look for alternative sources to supply the raw material. In particular lead, zinc, copper and nickel will face shortages between 2014 and 2030.[13]

In the paper industry the price of pulp from our trees also increased tremendously post-1995 and therefore more and more paper mills were gearing themselves to using waste paper (paper for recycling) as well as pulp. Eventually the processors realised that they could not go on taking in poor quality material which required further processing because their own costs kept rising. In other words we have not been supplying them correctly because of the contamination that results when recyclables are mixed with other material which should not have been there. If steel cans are mixed with copper that will result in contamination of the copper batch because the copper mill will not want steel in their loads. Likewise there is a big demand for aluminium drinks cans but these often get mixed with hairspray cans. Hairspray cans are a classic example of iron cans (ferrous metal). The aluminium mixed with ferrous is contaminating the batch so there has to be a separation of the two. With more and more material coming in the form of contaminated fibre (mixed ferrous and non-ferrous material), the end user is increasingly demanding 100 per cent clean material that can be

---

12   *The Skeptical Environmentalist*, Bjørn Lomborg 2001
13   *France Green Tech Magazine*

used directly. This emphasis on quality has increased dramatically in the last five years (to 2013). It should be acknowledged however that the most powerful driver here is the margin that the recycler expects to make, not necessarily, if at all, the more altruistic reaction to diminishing finite virgin resources. Recycling after all is a business and a transition is happening where the buyer not the seller of waste material is in charge. Nevertheless everyone recognises that we have to deal with the waste we create and if we are going to do something we might as well do it to the best of our ability.

So rather than ask ourselves: how do we recycle more, perhaps the question should be how do we recycle better? We should certainly be recycling more because around the world we have still not been able to reuse all the waste that is generated. It is estimated that global municipal solid waste generation levels alone are expected to increase to approximately 2.2 billion tonnes per year by 2025.

The ultimate challenge and the greatest risk is that we must protect the finite natural resources of our planet earth. For example there are different varieties of trees used to make pulp for paper production which accounts for 35 per cent of all felled trees. Recycling one tonne of paper saves up to 31 trees, 4,000 kWh of energy, 1.7 barrels (270 litres) of oil, 10.2 million Btu of energy, 26,000 litres of water and 3.5 cubic metres of landfill space.[14] Or to put it another way one tonne of pulp equates to about 357 reams of the photocopy paper that we use every day. If we relied on forests for all our pulp we would soon run out of land for agricultural purposes or to build hospitals, schools and homes. The World Wide Fund for Nature (WWF) claims that if we maintain current resource use we will need two planets by 2030! So obviously there is a limit to the amount of land available for forestry purposes even though we are now cultivating trees that can grow to full size in five years.

Today we are using some 400 million plus tonnes of paper per annum in a global population of about seven billion, but if we look at the anticipated population figure in 2045 of nine billion,[15] just on that basis alone paper production will need to increase by 30 per cent. On top of that within our current population not everybody is able to get a newspaper or has a book to read, or has enough exercise books to write on at school. As their percentage increases the demand for paper will also increase. We have limited land, limited number of trees and if we don't recycle we will not be able to supply the 400 million tonnes we need today or meet the demands of an increasing population. Remember of course that within the 400 million tonnes we use today about 200

---

14   Bureau of International Recycling (BIR)
15   In May 2011 the UN Populations Division raised its projections, estimating that there would be 9.3 billion people in 2050, and 10.1 billion at the end of the century.

million tonnes of waste paper is used in its manufacturing; 200 million tonnes of waste paper generates roughly 160 million tonnes of recycled paper. Therefore if we didn't recycle we would only have 240 million tonnes of paper so there would be a shortfall or we would have to use more pulp. What we already use today and take for granted would not be available. A sobering thought maybe when we glance down at our overflowing waste paper baskets.

There is another factor to add to the equation: the growing 'middle class' which grew by 55 per cent from 3.1 billion to 4.8 billion people between 1980 and 2009, representing some 71 per cent of the world's total population. Those in the high income class almost quadrupled earnings from $9,467 to $37,264 and grew to 1.1 billion people. If we stick with our paper example, the 400 million tonnes is meeting the requirements of the existing lower, middle and upper classes of today. When more people move into the middle class the demand for luxury paper and reading material increases; more people go to universities and the demand for new text books also increases. Even with the benefit of computers, iPads and Kindles which people can use as a substitute for paper-based reading the demand for paper will certainly increase. Just as the demand for reading and writing material increases so too does the demand for soft drinks which increasingly are packaged in cartons and the base product for those cartons is a form of paper.

While I have focused briefly here on paper the same principle applies to every other raw material which will be in ever-increasing demand as the world population grows and becomes more affluent and acquisitive. But I return to the question of quality. To save and recycle our products we have to give careful consideration to the first step in the process – the collection. Here the industry, the regulators and the local authorities cannot agree on this most elementary and yet essential link in the chain.

As I have suggested it is important, even vital, to keep one's eye on where the waste product is going to end up. I consider that the failure to understand this presents one of our greatest risks. As far as plastic and aluminium cans are concerned they can be treated and the aluminium cans will separate regardless of collection method, but paper is only of any use if it is in a dry state. The moment paper is contaminated with liquids, and some of these liquids can colour the fibre, then the end use for that paper is lost as a result and it can no longer be recycled for making paper. Single-stream or one bin collection points render paper unusable.

The idea of single-stream collection came about in the 1990s when there was a push to keep waste out of landfills and to increase collection rates as we have mentioned. But that drive was then superseded by multi-stream collection when there was a greater awareness about recovering valuable raw materials from the waste stream. Nevertheless, a lot of people are again in favour of single-stream collections, removing the easily recoverable material and putting the rest into waste-to-energy streams or Refuse Derived Fuel (RDF). It is proving very popular.

In Norway the citizens of Oslo enjoy the benefits of hot water heated by waste material which is burned. It is therefore essential for the operators to have as much waste as possible and as there isn't enough being generated in their own country the Norwegians are importing it. On the one hand this is an excellent way of providing relatively cheap energy but once all that waste has been burned it is gone forever. The system will demand more and more waste but how environmentally sound or cost effective is the practice? The priority in this case is actively promoting the one thing others are fighting – the creation of waste. Surely we should be cutting down on what we throw away.

The waste-to-energy lobby has a powerful argument: far better to burn waste than oil or coal. Very true but in the long run this strategy is ultimately destructive of finite resources. I know waste-to-energy is a good option because we cannot use 100 per cent of the waste that is generated and some of that goes to landfill, but RDF should only come into play after the maximum amount of recyclables has been removed from the waste stream. My worry is that the appetite of the machinery in Oslo and other cities using the same technology will be so voracious that the temptation will be to sell waste to them rather than take the tougher but more environmentally sustainable route of recycling. This brings us back to the competing priorities – the local authorities want to sell to the highest bidder, the highest bidding collector will have an eye on their profit margin and will undoubtedly be tempted by offers from those desperate to stoke their boilers.

Who is winning the argument? Whether it is waste-to-energy or recycling one of the key issues is, what are the add-on costs? If single-stream collection is used and provided the incinerators are close by, it could be argued that the cost aspect is reduced therefore the economics work. But as an environmentalist and supporter of the planet, I know that we must recycle the maximum possible amount from the waste that is generated before we look at any of these other alternatives. In my own city of London, we seem to be resorting too readily to incineration rather than investing in alternative recycling facilities. According to statistics released for 2012–2013 by the Department for Environment and Rural Affairs (DEFRA) almost half of the capital's 33 councils recorded a fall in household waste recycling rates while 40 per cent of what was collected was incinerated – an increase of 17 per cent over two years.[16] Attention is now focusing on shale gas and harnessing other forms of natural energies so we

16   *The Guardian* 27 November 2013

should not look at using waste for energy as the primary goal. As a secondary goal it has its merits but I certainly cannot back the proposition of using it as a first option.

Those in the waste industry would say they are only trying to come up with innovative solutions to a problem which is never going to disappear. We live in a throw-away society. Americans discard 40 per cent of the food they buy every year, worth $165 billion,[17] although it is worth pointing out that the average Mexican household throws away a third more than the average American home.[18]

With global Municipal Solid Waste (MSW) generation levels rising ever higher, what if anything can we do about consumer habits? First of all we haven't begun to scratch the surface in terms of collecting the waste we produce today let alone what lies ahead. Around the planet we have certainly not recovered all the possible waste streams available; there are countries in Africa, South America and Asia where the collection of waste from households has not even begun although collection of waste from industries has started in some form. But we in the developed world are consuming without thinking of the consequences; we are only adding to the problem. The challenge is to persuade every citizen on this planet only to consume what he or she needs. Before throwing a plastic bag into the waste stream we should first consider whether we can reuse it, then we will be reducing the burden on the recycling industry so it can concentrate on what one might call unavoidable waste rather than unnecessary waste. Before we order more food we should ask if we can we buy less. We need to look at all these aspects with a view to producing less organic and solid waste; we should be asking if we have had the maximum benefit from every product before we dispose of it.

Having taken the consumers and waste industry to task it is reasonable to ask what our governments are doing; are they hindering or helping? It is unfortunate that even within individual countries everyone is adopting different methods of collection particularly from households. Commercial waste has never been a problem; all we have to do is to encourage the supermarkets to reduce the waste that they create and to store it in a usable format. So a little more education at the commercial level will help us increase the recyclability of the waste they create, rescue that waste and make it more economically viable for the reprocessors so they get it in the right format.

The real challenge is the waste from the households, the community streams. Here there is no set formula; every authority resorts to their own systems and mechanisms or more likely to the systems that their waste partners demand (i.e.

---

17   National Resources Defense Council
18   Rathje and Murphy 1992, 216–19

the collectors). Different councils and authorities wherever they are in the world are governed by the partners they have. Depending on the type of MRF, the operators dictate how they would like to collect the waste. But at the very least the dry and the wet refuse must be collected separately. Because the volume of paper collected worldwide is so much higher than the volume of metal or plastic collected that is where we should focus much of our attention. We must look at the total amount of paper that can be collected worldwide; if this cannot be used then it will either end up in landfill or it will go to RDF purposes in waste-to-energy. But as I have explained waste-to-energy is not the answer because if we burn it we will never meet the demand for paper worldwide.

So, in reality who is actually calling the shots? The councils who are collecting this material go into a tendering procedure looking for companies to take the waste off their hands. In most instances it is the highest tender which wins the contract which is an entirely reasonable and commercially sound approach, but the highest bidder may or may not be the most eco-friendly of the waste operators and the council's responsibility does not end by simply offloading the material. The council's responsibility should be to continue to ensure that the waste that is collected from their communities is taken to a plant where it is reprocessed, sorted and delivered to the end user as raw material that is user friendly. Otherwise the quality of the material that is coming out of these plants does not really meet the requirement of the end user who will be less encouraged to buy the product. But local authorities have a duty of care to the planet as well as a commercial responsibility; after all they are a branch of government and therefore they have a moral duty to act sustainably, which must be considered in parallel with balancing the books.

I was recently asked at a conference where I put the blame for a drop in quality of recyclables: was I blaming the mills, the legislators or the recycling facilities? I pointed out that when I started in business over 30 years ago, the only serious problem we had with quality was contamination and this could be put down to the fact that we did not have very advanced sorting equipment, so occasionally we would have some lower quality printed paper mixed up with pure white paper. Other than that we didn't have any problems.

Sadly the blame today lies principally with the local authorities in every country which have ducked their responsibility for where the waste will end up. They are partners in collection and they are partners in making sure that the waste can be converted into a raw material which has got an identifiable market. They are 'partners in crime'. But what all the authorities have done now is put the waste out to tender to the highest bidder, washing their hands of it. The recycling plant gets the waste material which it depends on for its feedstock but because of the high tender price the recycler has paid leaves only a very small margin for processing and selling the recyclables. This selling price in turn is capped by, for example, the box manufacturer which in turn is capped by the packaging company that needs the boxes to sell the finished goods which in turn is capped by the final consumer price. The recycling plant therefore does not have enough margin for investment in better sorting systems which therefore results in what might be regarded as poor quality production. We should go back to the councils and remind them that their duty of care does not end by trying to get the highest price for their waste; in fact their duty of care continues down the chain in helping the organisation that paid the highest price by ensuring that the highest quality of recyclable is produced. If required the councils should consider funding the business and providing any infrastructure support that may be required as a joint public–private venture. The blame for our current quality dilemma in Europe can be attributed to the councils. Europe is considered the worst offender and yet we are the leading community waste collectors in the world.

Essentially the output of a recycling plant depends on the quality of the waste input. If the quality produced is not acceptable to the end users all they will do is move their custom to another authority, to another country or another continent. Therefore we find that because of the poor quality, many recycling plants in Europe have been losing export volume because they are not meeting the standards now required. This is a truly global business, where it does make sense to ship material half-way round the world and still make a profit. European reprocessors are looking as far as Japan, Australia, America, the Middle East and even Africa to ensure that they are getting good quality material, particularly commercial waste.

I mentioned briefly that the landfills of old are vast holes in the ground contaminating the soil around them. Now with advancing technology the quality of landfills is vastly improved. Dry or sealed landfills have sealed side walls so nothing can leach out and permeate the soil. Nevertheless even though hospital and toxic wastes are no longer dumped untreated into the ground, nobody wants a landfill near them. Not in my back yard (NIMBY) on any account. It is not so much that we have run out of landfill space, which we haven't, it is that authorities have radically reduced the number of licences for landfills they are granting.

One of the deterrents governments have used is an increase in the cost of using landfills – the landfill tax – so more and more people are economising on sending waste to them. In the UK the cost is about £72 per tonne (2013 prices) and likely to increase. In addition the cost of transport to the landfill pushes the cost up to £80 or £90 per tonne. The European Union is constantly setting tougher recycling targets to increase recycling rates and there are numerous and complex directives which emanate from Brussels governing how much and what we can simply dump. Despite the punitive costs landfills are still in use possibly because it is the easiest option – some estimate that 80 per cent of items buried in landfills could be recycled.[19]

Legislation is an important if complex and at times contradictory part of the waste industry because we don't want waste to go from one country to another unprocessed. Importing countries should only accept waste that meets specifications but because landfill taxes are rising it has been an easy solution just to ship untreated waste out of the country and dump it on someone else's doorstep.

But there are risks associated with an over-protective legislative programme which we are increasingly enduring today. In many instances we are in greater danger of being over-legislated rather than under-legislated. This not only impacts on the handling of waste but also leads to contradictions in the interpretation of these laws making it difficult to move waste from country to country; even within Europe for example laws handed down from Brussels are being interpreted completely differently by individual member states.

At the BIR's International Environment Council meeting in Warsaw in November 2013, the Chairman, Olivier François of NV Galloometal, pointed out that the French government had generated some 80,000 pages of new regulations in 2012, many of them relating to the environment. He observed that although mostly well-intentioned, such a volume of regulation was impossible to assimilate such that the interpretations adopted and the actions taken by enforcement agencies could not be predicted. As a result, the effects of this 'very bad situation' could be 'disastrous', he said.[20]

While legislation is important to control the quality of waste there is a real danger that it will impede the free movement of good quality material between countries. We seem to be caught up in bureaucratic exercises of filling in too many forms, writing too many descriptions and providing evidence on top of evidence for containers moving from A to B. Legislation is important and we will look at the whole physical and technical process of moving waste in a later chapter but the laws need to be streamlined and become more user friendly.

There is growing talk in the industry about closed loop recycling. It is a concept a lot of people support which says waste generated within the state or country should be collected, recycled or used and then sold within that state; in America this is known as 'trash independence'. It is supposed to be the most favoured form of responsible recycling – you live with the waste you create. I personally don't support this idea because every tonne that is recycled is part

19   Cheryl Jakab, *Global Issues: Clean Air and Water*. North Mankato, MN: Smart Apple Media 2007
20   BIR 4 November 2013

of an international closed loop and contributes to all our efforts to preserve declining resources.

In my closed loop there are no cross-border restrictions and every member is allowed to trade with anyone on the planet. To me the Earth is a single, homogeneous closed loop whether one is shipping waste to India or China from a plant in the UK for recycling before it gets shipped back to the UK for reuse or indeed to any other country that one chooses. The more highly restricted closed loop being proposed would be valid if we were able to consume all the waste that is generated within our own country, but we can't. In the UK we collect close to ten million tonnes of recovered fibre per annum and over the next five years (to 2018) that could increase by 20 per cent, but the total volume that Britain can use is just less than five million tonnes. If we were restricted to our own 'closed loop' what would we do with the other five million tonnes that we collect but cannot use? For me the entire ten million is part of a global closed loop programme and the purpose of that programme is to preserve our natural resources for the benefit of all mankind. We must encourage everyone to think like that because we are all members of the world's closed loop – we are in this together. At the same time we must remain competitive otherwise countries are going to become marginalised if buyers perceive that they are not getting the same treatment as others within the closed loop.

A better definition of the closed loop is a programme promoting recycling within a loop which is defined as generation of waste and transportation to a plant that sorts it into a recyclable format. This is then transferred to a reprocessing partner who could be anywhere in the world and the finished goods come back to a manufacturer whose products can be sold anywhere else in the world. That to me is a closed loop. Another definition of a closed loop is one in which, say, a tin can is constantly recycled and never ends up in landfill – no one would disagree with that. As always it comes down to economics. In America 'trade in trash' is worth some $4 billion per annum[21] so it is unlikely that 'trash independence' will prevail.

So much attention is devoted to how we treat waste that too often the important element of logistics is neglected, which to me as a trader is vitally important in the whole chain. We will look at this aspect of the recycling industry in more detail but in outline the movement of waste is critical.

It seems obvious that we are generating so much waste that we need to transport it around the world. But before the waste is ready to be exported we have already incurred costs in terms of collecting the waste, moving it to a plant where it can be sorted ready for reprocessing at another facility. The reprocessor

---

21   *Recycling Myths Revisited* by Daniel K. Benjamin

dictates the price and in turn is governed by the amount, say, the box maker is paying for cardboard and packaging. Therefore the price is always governed by the end product price which is influenced by many other factors including the price of virgin raw material. These are the challenges that are faced on a daily basis. We have to keep a constant eye on the cost of transport within a country, the export cost of shipping a container load of waste and the comparative price of the product from our competing countries. Globally in the paper industry alone there are some 40 million tonnes plus of waste paper traded between the continents annually. Add into that mix all the other waste streams and it should come as no surprise that close to one million tonnes of recyclable raw material is on the high seas every week of the year. *Forbes* magazine described the USA's traffic in waste colourfully:

> Our trade to China is, in great part, about sending raw hides and getting back shoes; sending waste paper and getting back packaged toys, lamps and household items; sending scrap metal and getting back machinery; and sending raw cotton and getting back finished clothes. Wal-Mart Stores alone imports 576,000 containers from China each year.[22]

One last consideration in this short overview should be of the key players. In Europe and America a handful of companies tend to dominate and they are beginning to turn their eyes on the Asian market which is relatively free from this sort of monopoly position. While it is important to have strong global players taking on the European and US challenges of handling and processing waste, they must not be allowed to endanger the growth of small and medium sized enterprises (SMEs). Local authorities should also become friendlier to promoting SMEs in handling the waste. Here again quality is the main issue. The councils not only have a responsibility to dispose of the waste from the community channels but they also have a bigger responsibility of ensuring that the waste coming out of the community streams ends up in the right quality so it can be used in the most environmentally advantageous way by MRFs. It should not just be processed for the sake of processing, producing an inferior quality which is not wanted by the global market. Local authorities have an important role to ensure that the partners they choose can also become their partners in guaranteeing that refined recyclates are produced even if that means reinvesting some of the money they get from tendering the waste. Big players have an influential role but they should not be allowed to work against SMEs and benefit from a favourable tendering process which makes it difficult for SMEs to participate.

---

22 *Forbes* 24 May 2006

In short the parameters for tendering should be more SME friendly which would enable more players to participate together to produce better quality waste, sorted in the most user friendly format, while working on a zero tolerance to landfill strategy and keeping it cost effective rather than volume effective. Too many MRFs are just running on a volume principle meaning we have volume in and garbage out. We want garbage in and quality out.

To summarise, in one sense as an industry we should be optimistic because we have a raw material – waste – which will never run out unlike the finite virgin resources which are declining or are too costly to exploit. However it is proving difficult to keep pace with the demand of a growing and increasingly affluent population. It is no use clinging to the hope that the rate of population growth is slowing and that in half the countries of the world we are only having just enough children to replace an ageing population. There is a stark warning from the United Nations:

> The latest UN revision recognised that family sizes in Africa were not falling as projected. The model predicts that Africa, now with 1bn people, will have 3.6bn by 2100. In Niger, where the rate of population growth has outstripped economic growth for part of the last five years, a fifth of women have ten or more children and a third of children are malnourished. If the birth rate remains at current levels, the population will soar from 14m to 80m by 2050; even if the rate is halved, there will still be a projected 53m.[23]

I would say we have no choice; our finite resources are limited and by depleting them further we will destroy our natural environment. Our only option is to recycle what we use, but we must start getting cleverer about how we achieve our ends, using a long-term strategy rather than seeking short-term solutions, by improving the quality of our recyclable material and by maintaining a clear focus on the end product, supported not hindered by legislation.

23   *Prospect Magazine*, 2013

2

Myths, misunderstandings and mistaken targets

Like many other industries, the waste sector has its fair share of sceptics and doubters about our techniques and our motives. Buzz words and catch phrases abound as we talk blithely about zero waste strategies and closed loops, and we are not immune to jumping on the newest idea in the hope of solving what simply may be insoluble. The danger for all of us is that we succumb to any one of these half-truths and base the whole of our policy on that criterion. Recycling cannot be taken out of context and it should not be tackled in isolation. The waste we create and the methods we use for dealing with it are part of a wider picture which is constantly changing. We are part of the renewable energy world and part of the preservation of our natural environment; we consume energy to produce our products, we consume energy in the process of disposing of our waste and our industry is inextricably linked with finding ways to generate new energy. Waste prevention and recycling are integral in reducing and preventing climate change.

When we dump our waste in a landfill that is not the end of it. Landfills, which will always have a role to play, are living entities producing a complex mix of gases as a result of the chemical reactions taking place in the waste. There are benefits. In modern landfills the bacteria break down the waste and a landfill gas – half methane and half carbon dioxide – is produced. The gas can be captured, vented or burned, or in the older ones just allowed to escape into the atmosphere. It has an impact on global warming. In the United States, landfills are the largest anthropogenic emitters of methane gas – in 2006 landfills there released 6,211 tonnes or 34 per cent of the total emissions, although the amount released into the atmosphere actually decreased between 1990 and 2007 because the gas was collected. [24]

---

24   Greenblue 2011

Anaerobic digestion, the process by which microorganisms break down bio-degradable material, is a valuable source of renewable energy. Poorly managed 'dumps' are destructive to the environment; professionally managed landfills are less so – nothing is neutral. We are rightly concerned about the effects of green-house gases on our planet but landfills have a role and we should not create artificial barriers for ourselves in our headlong pursuit to meet well-intentioned targets.

There is disagreement among scientists about the causes of global warming. In September 2013 the UN Intergovernmental Panel on Climate Change pub-lished its fifth major assessment of what was happening to the planet, warning about the changing circulation and temperature of the Atlantic and the Gulf Stream and the impact of carbon emissions. They were 95 per cent certain that humans were responsible for global warming – a figure which had jumped from 90 per cent in an earlier report – and called for 'substantial and sustained reduc-tions of greenhouse gas emissions'. But people are both weary and wary of dire predictions. One newspaper commented: 'Some will warn that the report gro-tesquely underestimates the dangers of imminent global apocalypse; others that the whole thing is a tissue of fabrications from an organisation that has lost any credibility.'[25] We are told that the Gulf Stream which influences our climate was being affected by increased carbon emissions and yet Professor Carl Wunsch, Professor of Physical Oceanography at the Massachusetts Institute of Technol-ogy, says in fact it is predominantly driven by the wind and has always urged caution when jumping to conclusions. While accepting that global warming is real and has a human-made element he said in 2007: '…the notion that the Gulf Stream would or could "shut off" or that with global warming Britain would go into a "new ice age" are either scientifically impossible or so unlikely as to threaten our credibility as a scientific discipline if we proclaim their reality.'[26]

This is not the place to fight the global warming debate but I would say the counter-arguments all need to be scientifically not emotionally evaluated by our politicians and legislators.

The pursuit of alternatives to burning fossil fuels is laudable but there are consequences which we sometimes choose to ignore. Perhaps one of the most obvious examples is the proliferation of wind turbines. I shall avoid the increasingly hostile debate about their efficacy[27] and impact on the countryside,

---

25   *Daily Telegraph* 27 September 2013

26   http://www.realclimate.org/index.php/archives/2007/03/swindled-carl-wunsch-responds/comment-page-3/

27   Dr John Constable of the Renewable Energy Foundation was quoted as saying: 'Wind energy is an experiment, and sometimes the lessons learnt are hard and dearly bought. The truth is that foolishly ambitious targets and silly levels of subsidy have overheated the wind industry, resulting in defective technologies and poor installations.' Councils in the UK were said to be generating as little as £13 worth of energy a month from the wind turbines (*Daily Telegraph* 27 December 2013)

but they are not as clean as people might imagine as they are creating their own toxic waste dumps which are beyond help from any recycling programme.

As turbine blades turn leisurely in the wind, inside their motors – they are not silent – are powerful magnets helping power the generators and these magnets typically require 4,400 lbs (1,995 kg) of neodymium-based material. Neodymium is one of those rare earth metals which can be found in remote corners of the world but to extract it requires a poisonous process resulting in millions of tonnes of radioactive waste being dumped every year in tailing lakes – the runoff from the extraction. This toxicity can lead to serious side effects such as teeth beginning to fall out, hair turning white at unusually young ages, and people suffering from severe skin and respiratory diseases. Children are born with soft bones and cancer rates are increasing. The question for the supporters of wind turbines is: is this a price worth paying?

Alternative energy has its place but there are consequences and while targets to find alternatives to fossil fuels and other finite resources continue I fear targets are being plucked from the air by those without a detailed understanding of the waste industry in a headlong pursuit of some arbitrary percentage. Let us consider just some of the other myths and misunderstandings.

It is often said that we are producing so much waste that we will soon run out of landfill space. That is obviously nonsense in some areas. The United States has more than enough land to dump all its waste for centuries to come but some smaller emerging countries such as those in the Far East are not so fortunate. They have got to find alternatives to burying their waste. Once they have followed the principle of reducing what they produce, reusing anything they can, they turn to recycling but what should they do when a waste product has reached the end of its useful life? If they cannot put it into landfill it either has to be disposed of in RDF facilities or traded internationally to other members of the global loop who can use it as raw material.

What is happening in 'developed' nations is not that they have run out of landfill space but that new licences are not being granted – in Switzerland they have banned them. So there are fewer landfills and while the new ones are bigger there is a reluctance to use that method; in addition new legislation is actively preventing dependence on the landfill option and landfill taxes are rising.

Once again the legislators may be missing the point: we are becoming too target driven rather than product driven. The emphasis is on a continuing reduction of the use of landfills almost as an end in itself without consideration of the real implications of that directive. If landfill usage is restricted and there is an ever-increasing list of rules and regulations, in particular those defining the difference between when refuse reaches the end of its useful life and what is recyclable product, it will in time cause a bottleneck in the whole process of trading recyclable products around the world.

It may be worth briefly explaining here how a modern landfill operates. The bottom is lined to prevent leakage into surrounding soil or groundwater, there are a series of cells to hold the waste which are capped at the end of every day, there are drainage systems to collect rainfall and as I have mentioned sometimes a process of capturing the methane gas. That of course is the ideal but there are many examples of 'dumps' sometimes of highly toxic waste; the former is desirable, I would say a necessity, the latter of course should be condemned.

One of the consequences of simply saying no to any more landfills will be an increase in the amount of illegal shipments of toxic and contaminated waste to countries which are prepared to accept them and then simply dump them. It is thought that the market value of illegal waste shipments was between $10 and 12 billion every year but is now reducing due to tighter controls. The European Commission estimates that 25 per cent of all waste shipments from the European Union are illegal and contained in those shipments will be mixed waste much of which ordinary householders may well have thought they had carefully sorted. Unscrupulous operators are trying to dodge the regulations, many of which are essential, but unless legislators can see the difference between targets for their own sake and the quality of what is required for the end product then we will have an increase in illicit cargoes.

Incinerators are also roundly condemned for being polluting and dangerous to health particularly if one is to be built in your neighbourhood. The old ones undoubtedly were but today there is a strict enforcement of emissions and an EU Report in 2006 considered that pollution risks were generally very low and in all probability over the next 10–15 years there is likely to be an increase in the number of MSW incinerators driven in large part again by local authorities and cities having to comply with Landfill Directives. We are chasing targets.

If we accept that the very latest and best incinerators are safe – and I recognise that there are those who will never accept that – then we should consider whether they are cost effective and that depends on where they are situated. If they are close to a MRF, which means transport costs can be kept low, then there is a reasonable chance that they will be; however it may not be long before they run out of suitable feedstock – the waste that is burned. As we noted in the previous chapter cities such as Oslo even with affluent consumers already have to import waste to keep the fires burning but in time the sums may not add up. I know it sounds extraordinary but some waste is scarce. Even today we see many good, modern MRFs having to be mothballed because of lack of good feedstock.

The reality is that it will require a carefully structured balance of incinerators,

21

WASTE PREVENTION

RE-USE

RECYCLE - COMPOST

ENERGY - RECOVERY

DISPOSAL
(LAND)

Figure 2

MRFs and landfills to make the process work. The pyramid is illustrated in Figure 2 with the emphasis of course on the prevention of waste in the first place.

One of the biggest targets for abuse is plastic. Polyethylene is the most common form of plastic and we produce 80 million tonnes of it every year.[28] Ethylene is derived from natural gas and petroleum. We use too much of it, wrap everything in it and then just throw it away. In fact the plastics industry has been working hard to address much of this. In the first place we as consumers were all too ready to embrace plastic for every reason from hygiene to convenience when it first arrived, so we should not rush to blame the supermarkets. Strangely enough by using plastic and polystyrene packaging we are in fact keeping waste down as it allows us to buy what we want (although we still buy too much and throw a lot away): for example, buying four chicken breasts instead of two whole chickens means we are cutting down on what we would inevitably dispose of in our kitchens. And yet in October 2013 Tesco, a leading supermarket chain, announced the results of its own survey of its stores and distribution centres in the UK for the first six months of the year and discovered that between the company and its customers nearly 30,000 tonnes of food had been thrown away. There is clearly more that can be done.

---

28   Piringer & Baner 2008

The cry is: well if we are going to use so much plastic, it should be biodegradable, that way it won't fill up our landfills. Actually it will. Biodegradable plastic is thicker than ordinary plastic and when it 'degrades' into landfills it does not disappear; indeed in volume terms it takes up more space albeit in tiny pieces. The industry has been trying to resolve the problem by changing the structure of the plastic used in such things as supermarket bags and looking at producing polyethylene from everything from sugar cane to sugar beet and wheat grain so when they are disposed of they do cause less environmental damage. We should also remember the additional processing costs of producing biodegradable plastic – the raw materials used and the energy expenditure. Are we encouraging the use of these bags simply because they degrade or should we be focusing on using fewer bags in the first place? There is an important role for supermarkets here which we will consider later but they of course see a business opportunity with all the waste that is produced and now increasingly provide their own convenient waste bins outside their stores because they know all that rubbish has a value.

Plastic has an undeserved reputation for being a major component of landfills. In reality it is not plastic which is filling up all those collecting bins, landfills or even the containers sailing round the world. It is paper. In the UK the ten million tonnes of paper collected every year compares with less than one million tonnes of plastic and the same proportions, I would estimate, apply worldwide.

In fact there is worldwide demand for plastic scrap of about 42 million tonnes (current market value about US$17 billion) and that figure is expected to rise to 85 million tonnes valued at about US$40 billion by 2020. A substantial quantity of plastics scrap will be generated following the introduction of European rules and regulations relating to automobiles and electric and electronic wastes. Europe only recycles 26 per cent of current plastics scrap generation so there is some way to go.

While we moan about all the paper we use and decry our ubiquitous plastic bags we wouldn't be without either; therefore the issue is how do we deal with it all, how do we stop building the waste mountains we roundly condemn?

3

Industry operators, systems and capacity

The leading operators in Europe are the French-owned Veolia and Suez Environnement followed by Remondis of Germany and FCC of Spain. In the UK, Biffa, which was acquired by a buy-out company backed by Montague Private Equity and GE Capital for £1.23 billion, heads the British list – the sector is estimated to be worth £11 billion employing 100,000 people. In the United States the leaders are Waste Management Inc., Republic Services Inc., and Veolia Environmental Services North America Corp. Waste Management enjoys revenues of around $12 billion. This is the rarefied world which SMEs can only dream about, but let us take a step back because it is also a world which is slow to have an impact in the areas of greatest need.

As the volume of waste started growing it immediately raised two major problems for the market: what to do with the waste and where should this waste be treated and recycled? So the cry for help went out certainly in Europe. The smaller SMEs could only take a limited share of this mountain of waste, based on the capacity of the plant they were operating. If they were processing 50,000 tonnes a year they might have been able to ramp it up to 60,000 but they certainly could not manage 100,000 tonnes. Nor did these SMEs have the funding to build an additional line to increase the volume that they could handle.

What this really meant was the industry needed outside support. The visionary business leaders led by medium and large financial houses who partnered with them helped in building a number of substantial facilities and each one has grown through acquisition to become one of the top 20 companies in the Western world.

The consequence of this consolidation has been a broad-based monopoly among the elite group. As these companies have grown, their market share in their own right has increased and now they control more than 50 per cent of the market in Europe. There is an appetite for growth and these companies are constantly developing new areas of activity in metal, plastics, textiles and construction waste not to mention all the latest waste-to-energy technologies. They are evolving and at the same time shedding the less profitable side of their operations as well as looking to diversify in utilities such as water.

The downside of this is the lack of competition and growth by SMEs. I would like to see SMEs getting the opportunity to participate in this expanding market. But the competition in bidding for that opportunity and the kind of investment required to set up or expand a plant is too high. The challenge in bidding for new contracts is not bringing to the forefront as many SMEs as one would expect. Many SMEs find it difficult to participate in the tendering process because the tender conditions or performance bonds or guarantees may make it too difficult for SMEs to enter the bidding process. These kinds of obstacles are constantly working against them which can have serious consequences for their future viability.

Even the major operators struggled during the recession following the collapse of the financial markets in 2008 and were forced to cut the weakest branches of their business and lay people off. But as those smaller recycling facilities are mothballed it leaves a gap in the whole recycling programme for their particular region. Once again logistics comes into play – just how far can collection and recycling companies travel for their feedstock while still making economic sense? The answer is not far as the margins are so tight and it also depends on the region. In America, Waste Management found it cost effective to close some of its smaller facilities and transport the waste 100 miles rather than maintain a fully manned facility handling smaller volumes per annum. This may not add up in some European countries where we are already struggling with under capacity. The UK Government's Waste Resource Action Programme suggested that Britain alone will need 450 waste composting plants by 2020 just to meet the requirements of local authorities.[29]

And what is the impact on the emerging markets where there should be a greater focus? As noted above, the world population is growing; but not only is it growing it is changing in its habits. Between 2005 and 2025 it is estimated that food waste production in Asia is likely to increase from 252 million tonnes to 377 million tonnes. The greater affluence also leads to the consumption of more electronics and computers so while waste technology may be constantly reinventing itself at the highest level to cope with the ever-changing waste stream, what is it doing for the less privileged?

Irrespective of the type and source of waste that is collected it all needs to be processed. First let us look at some basic systems.

---

29 *Investing in Waste & Energy Resources* by Nick Hanna (Harriman House)

There are generally three types of MRF: clean and dirty (both used for dry material) and the wet facilities. This Wikipedia description puts it simply:

> A clean MRF accepts recyclable commingled materials that have already been separated at the source from municipal solid waste generated by either residential or commercial sources. There are a variety of clean MRFs. The most common are single stream where all recyclable material is mixed, or dua-stream MRFs where source-separated recyclables are delivered in a mixed container system
>
> A dirty MRF accepts a mixed solid waste stream and then proceeds to separate out designated recyclable materials through a combination of manual and mechanical sorting. The sorted recyclable materials may undergo further processing required to meet technical specifications established by end-markets while the balance of the mixed waste stream is sent to a disposal facility such as a landfill.
>
> A wet MRF uses mechanical biological treatment technologies to separate and clean output streams. It also hydrocrushes and dissolves biodegradable organics in solution to make them suitable for anaerobic digestion.[30]

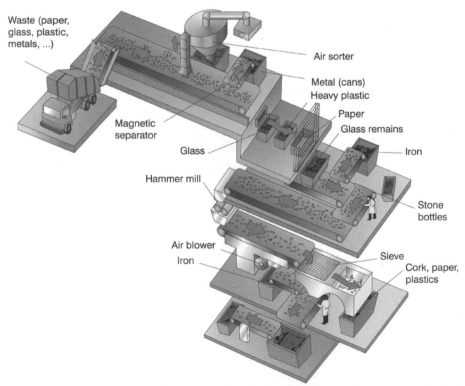

Figure 3 A dirty MRF (Source: http://en.wikipedia.org/wiki/Materials_recovery_facility)

---

30   http://en.wikipedia.org/wiki/Materials_recovery_facility

A very good MRF today would receive community waste in bags which would be ripped open by a bag splitter. The material would then be transported along a conveyor belt through a multi-sorting process. The objective of this process would normally be removal of wood or other large contamination to prevent damage to the machinery. The waste might then go into a drum to break up the waste and segregate small and large particles. As the larger pieces continue along the line glass fragments are removed because they cannot be processed with the paper and plastic. There is a mechanical process to pick up larger items such as sheets of cardboard which forms a good percentage of the waste, plastic bags are blown by air and separated while optical sorters and computerised separators help to sort the different colours of plastic. Magnetic separators will pick up the metallic items and an electrical process will isolate aluminium cans. So depending on the level of investment in MRFs one can achieve a high level of recovery from the waste. There is a lot of variation depending on the degree of investment and number of lines there are for sorting the waste. Some sorting stations may simply sort and bale aluminium cans while at the other end of the spectrum they might have a highly sophisticated and expensive plastics washing, sorting, shredding and preparation system.

Poorer emerging nations cannot afford such sophisticated technology and to a great extent they are being neglected; although they are continually being tempted to 'invest' in the latest facilities when that may not be their best option given the low or even non-existent level of their commercial and municipal collections. Modern sorting systems allow us to sort mechanically by colour, by type and by weight. It costs a great deal but it produces a product which is usable – a raw material. In these countries waste consumption or waste production per capita is much less than the waste created in Western countries therefore they cannot justify putting in an expensive plant and yet it is in these regions that waste is growing and where the least recycling is carried out.

Initially these emerging countries should consider primary sorting rather than the secondary level of sorting. At the very least they should be trying to segregate dry and wet waste and thereafter consider sorting the dry waste into paper, plastic, glass and metal.

Scavenging on the waste dumps in Africa and India is a livelihood for so many; even their homes, often no more than shacks made out of scraps of wood and tin, are built on these sites; but how much is actually going to be collected and how much ends up in landfill or polluting the rivers? How many toxic substances hidden below the surface are spreading out and possibly affecting crops as far as a couple of miles from the dumps?

In these emerging markets it is important that we have proper controlled methods of collecting, sorting and delivering recyclables in a format that not only helps the economy but also provides healthy employment locally. This is where the growth will be in the next decade or so, not in eking out that last percentage of recycling in the suburbs of London or the affluent arrondissements of Paris.

The waste management market is a $1 trillion industry which could double by 2020 and the biggest opportunity has to be in the emerging countries of Asia, Africa and South America. Analysts at Bank of America Merrill Lynch say this is the next big investment boom. 'We see opportunities across waste management, industrial treatment, waste-to-energy, wastewater and sewage, E&C [engineering and consulting], recycling, and sustainable packaging among other areas', they said.[31] So with proper focus this should be a golden opportunity, maybe even the industry's equivalent of the dot.com boom.

The systems therefore exist but not all are suitable for every nation. Appropriate rather than expensive solutions are the answer for emerging nations but in the West capacity is the most important factor. In Europe in particular we should concentrate our efforts on ensuring we have sufficient facilities to handle the expanding waste streams because populations are growing in developed countries too, creating ever-changing problems as people migrate in search of new opportunities. According to Eurostat Britain had Europe's fastest growing population in 2012 with 392,000 more people in the country bringing the total to 63,888,000 people. We do not have enough recycling facilities to cope with those numbers so where is the waste going?

There is a growing call for European waste to be kept and recycled in Europe – the principle of 'proximity': the so-called trash independence we have touched on in America. At a trade fair in Rimini the general director of Assocarta, the association of Italy's pulp, paper and board manufacturers, Massimo Medugno, said: 'A solid industrial structure like the recycling chain needs to be strengthened by implementing the principle of proximity by law.'[32] But as we cannot handle all the waste paper we produce in Europe at the moment it prompts the question: how much landfill will be needed to cope with the additional 12 million tonnes of paper we currently export? This will not please the environmentalists nor will it help reduce our landfill quotas. I am firmly of the belief that there must be no restrictions on the free movement of properly controlled safe waste.

There is one other possibility which is now being discussed suggesting that the whole concept of waste collection in bins outside our homes and business premises will change completely. Automation will be the key, potentially helping us handle excess capacity.

The futurists – although perhaps not as distant as some may believe – consider the process of transporting mixed waste for miles to waste dumps to

31 Emerging Markets 11 April 2013
32 PPI Europe, 21 November 2013

be archaic not to say unhealthy and uneconomic. The image of someone piling waste into a plastic dustbin and perhaps having to wheel it to the end of the road where a dustcart comes to collect it is somewhat old school in our world of slick technology. The modern apartment in the rapidly growing high-tech cities in those countries in Asia where the explosion of waste is happening may very soon have built-in systems in their kitchens to capture the waste, shred it and even analyse it.

A joint report by Veolia Environnement and the London School of Economics discusses the concept of nanoscopic robots which will analyse everything we throw away.[33] By 2050 cities will be more efficient, cutting emissions by 80 per cent and we will have 'intelligent packaging' with additives which would self-degrade after a few weeks or even years. According to the report, *Imagine 2050*, the packaging might even change colour depending on storage temperature to indicate when it is no longer safe to consume. This would do away with sell-by dates and cut down on waste.

All this technology apparently exists at least in experimental format so perhaps we will after all be able to cope with our population explosion but in the meantime we will still have to fill our multi-coloured bins and have them carted away every week.

While we are blessed with ingenious engineers working on every aspect of our industry we still have basic questions to answer and they relate to consistency and coordination of effort, an ability to see how rules affect every element of the recycling programme from product creation to its safe disposal at the end of its life. Somehow we have to find a way of including small operators because they take care of the waste stream at its lowest but equally important level. When that waste recycling centre closes in a village the impact is spread for miles around. The public already finds the whole process irksome even though they recognise its importance. We must not put up more barriers. It is evident that we know how to deal with waste in all its shapes and sizes; now we have to cooperate while we compete to ensure the public is with us. Operators can buy up and discard SMEs as it suits their business plan but while that strategy may help a company's bottom line it will always result in more waste being discarded irresponsibly.

Above all, the handling and treatment of waste has to be a partnership between governments and industry. This very issue of irresponsibility and worse still crime was raised in the UK when a number of the nation's representative bodies wrote to the Resource Minister, Dan Rogerson,[34] saying: 'We hope that

---

33   *Waste Management World* 22 November 2013
34   The statement and letter were signed by Anaerobic Digestion and Biogas Association, the Chartered Institute of Wastes Management, the Environmental Services Association, the Renewable Energy Association and the Resource Association, 22 November 2013

cracking down on waste crime will remain a priority for DEFRA, and that…
with cuts in Environment Agency staff sufficient resources will be allocated to
tackling this issue.' Seeking to work with the Government rather than criticise,
the organisations pointed out that 'the rate of increase in household waste
recycling and composting in England has levelled off and it is not certain that
the EU target of 50 per cent by 2020 will be reached'. They noted that 'DEFRA's
2010 Strategy on hazardous waste has not yet been implemented', they remained
worried about the confusion surrounding the interpretation of basic waste
collection and stressed that there was an infrastructure gap for commercial and
industrial waste. Capacity therefore remains an issue in the UK at least and it
is likely that the same situation prevails in other countries and markets around
the world.

4

Industry drivers

I am worried about the direction in which the recycling industry is heading. We know waste will never disappear indeed it will only get worse or from another perspective will, for the foreseeable future, be in great demand. So inevitably we have to have rules and regulations about how to tackle it in the best interests of the environment, the economy and even our health. Waste is a valuable raw material which we should not squander but it can also be hazardous; the politicians and the lawmakers have had to intervene in this dilemma to provide some workable framework. Unfortunately there is no general agreement over such legislation, not even about the most basic definitions of 'when waste ceases to be waste', which is known as End of Waste legislation. Furthermore every nation rightly wants to protect its own interests and those interests range from doing the very best they can for the national environment aiming for the zero waste dream to being quite happy to profit from unscrupulous operators importing the most dangerous waste regardless of the impact on their citizens.

The drivers which propel the recycling industry are regulatory, industry, supply, market and, overshadowing them all, the concern about the depletion of our finite resources.

Regulatory. The most obvious regulation is the landfill tax which in essence is a tax on the disposal of waste designed to encourage waste producers to find alternative methods of disposing of waste such as recycling or composting. When the community waste collection programmes started, more and more waste was collected and it had to be treated; in the early 1980s we found ourselves facing mountains of waste paper. In Germany when they introduced regulatory controls they launched what was called the Green Dot Programme – the 'Grüner Punkt' in the 1990s. It was binding on all businesses that if they used packaging, they were responsible for recovering their own packaging. Similar systems were soon rolled out in other European countries.

Landfill tax applies to all waste disposed of in a landfill unless the waste is specifically exempt. The aim is to break the link between economic growth and waste growth so that most products should be either reused or their materials recycled. Failing that energy should be recovered from other wastes where possible which ideally leaves a small amount of residual material for landfill. This preference order of waste treatment has been labelled the waste hierarchy.

The various laws and directives have proved effective. In the UK for example the proportion of waste deposited onto/into land decreased by 11 per cent between 2004 and 2008 (from 171 million tonnes to 152 million tonnes). In contrast, the quantity of waste recovered of all grades has increased by more than 50 per cent, from 95 million tonnes in 2004 to 142 million tonnes in 2008. The aim of course is to do better; the revised EU Waste Framework Directive sets a recycling target of 50 per cent for household waste by 2020. In comparison, Germany already recycles over 70 per cent of its household waste. And it is largely the landfill tax which is driving this change. In the UK it was set to increase from £72 per tonne in 2013 to £80 per tonne in 2014.

Figure 4 UK landfill tax, 2008–2014 (£ per tonne)

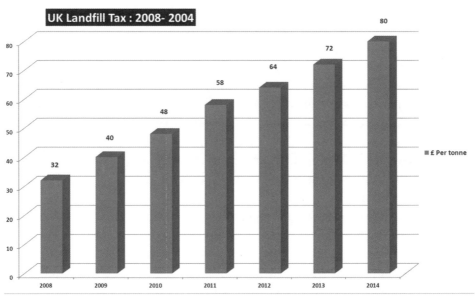

The landfill tax had to be increased to provide a disincentive for the prevailing habit of dumping everything in a hole. As a result people started realising that it cost too much to dispose of their waste in the ground, which helped promote greater recycling, which is where the recycling industry comes into the picture;

sorting the waste paper, the organic waste, the tins, the plastic and the glass separately so everyone is able to get value from waste that would otherwise have been lost forever in a landfill. If *high-density polyethylene* (HDPE) plastic milk containers were not recycled then the price of virgin HDPE would become so high that the cost of milk would rise or we would soon have to buy our milk in some other form of cheaper containers.

Recycling makes economic sense: we are able to buy products more cheaply, we are extending the shelf life of finite resources and we are reducing the emissions of greenhouse gas; in short there are multiple benefits beyond the landfill issue.

Figure 5 Timeline of significant regulatory events (list not exhaustive)

	EU Landfill Directive 1999/31/EC	EU Decision 2000/532/EC	EU Regulation 1013/52006	Waste Framework Directive (WFD) 2006/98/EC	Revised Waste Directive 2008/98/EC	Waste Directive Promotion/ Enforcement
EU Level	• Issued by EU and imposed on EU member states. • To prevent or reduce as far as possible negative effects on the environment and health from landfilling of waste, during the whole life-cycle of the landfill.	• Established a list of wastes and types.	• Established regulation for the shipment of waste.	• Directive 2006/12/ EC replaced Council Directive 75/442/EEC of 15 July 1975 on waste. • Significant amendments on definition of waste, waste prevention, recycling and processing of waste.	• Set the basic concepts and definitions related to waste management and lays down waste management principles such as the 'polluter pays principle'. • Set waste hierarchy.	• E.g. European Week of Waste Reduction 19-27 November 2011 (EWWR) • 2.672 EWWR actions were implimented in 2009. • 4.346 EWWR awareness raising actions implemented in 2010.

Industry. As the recycling industry develops to meet the new laws a number of different systems have been introduced to collect and treat waste including drop-off centres, buy-back centres and kerbside collections and within those systems there will be mixed waste collection – everything into one bin, co-mingled – a separation of dry and wet/food waste, and full separation. To meet recycling targets in the UK, local authorities have turned to mixed and co-mingled waste collection methodologies which require the use of materials recovery facilities (MRFs). This has put pressure on the current MRF capacity. Given the UK's high kerbside recycling rates, the nationwide MRF recycling capacity (in 2006 estimated at 2.5 million tonnes per annum representing the combined tonnage of 82 facilities) will soon be filled because of significant growth in the collection of dry recyclables, driven by changes in collection trends and increased consumer participation in kerbside schemes. Unfortunately it is not just the capacity but the available technology in these MRFs which is an issue and herein lies the problem which is facing many countries – vast quantities of waste are being recycled and not all of it can find a market. We will return to this later in the chapter.

Supply. The lack and cost of finite virgin resources means the only way to fulfil our production needs is to recycle our waste or consume those precious resources. It is generally acknowledged that consuming finite resources which are limited is destructive to our environment while recycling is more eco-friendly as well as being cheaper. It takes up to 24 trees to produce 1 tonne of paper using the Kraft chemical pulping process[35] and recycling 1 tonne of paper saves 2 tonnes of wood. WWF advocates zero net deforestation and forest degradation (ZNDD) by 2020 as a target that reflects the scale and urgency with which threats to the world's forests and climate need to be tackled.

It has been estimated that recycling half the world's paper would avoid the harvesting of 20 million acres (81,000 km^2) of forestland. While forestry statistics vary widely, there is consensus that consumption of paper has grown 400 per cent in the last 40 years. In 2007, it was reported that nearly four billion trees worldwide are cut down each year for paper, representing about 35 per cent of all harvested trees at the time.

If we didn't recycle then the cost of paper would be higher than current prices. If we didn't recycle the cost of the Coca-Cola we are drinking might be more because all the aluminium would have to come from bauxite. If we didn't recycle, the long-term supply of bauxite to make aluminium, the long-term supply of pulp to make paper, the long-term supply of iron ore to make steel and iron would probably have a limited shelf life. We have to extend the life of all those finite resources; therefore recycling is a necessity – an economic necessity and environmental necessity not simply an argument to cut down on landfill.

Running out of natural resources is not an idle threat. In China the National Development Reform Commission has officially designated 69 cities as being resource-exhausted; cities like Gejiu in Southwest China's Yunnan province, site of 24 per cent of the world's tin deposits first discovered during the Eastern Han Dynasty (AD 25–220). No more. The last 200,000 tonnes of tin deposits are forecast to run out in 20 years at the most. Continuous mining has not only plundered the supply but caused metal pollution and desertification. But the authorities are now fighting back. They have poured in state aid and through recycling turned the iron waste into an annual $654 million industry.[36]

Market. Waste is an internationally traded commodity and paper heads the list. Pulp and paper can be traded by global derivatives brokerage firms offering both over-the-counter (OTC) and exchange-traded transaction execution future listed on the Chicago Mercantile Exchange (CME).

By its nature only a fraction of the consumed paper material is recycled which, together with increased demand, implies that the need for recyclate material is always greater than the global supply. As a result, waste has become

---

35 A process that converts wood into wood pulp by treating the wood with a mixture of sodium hydroxide and sodium sulphide

36 *China Daily/Daily Telegraph* 23 October 2013

an international tradable commodity. For example, recycled white letter paper traded at £120–140 per tonne in January 2009, versus £160–170 per tonne in January 2010, and £240–260 per tonne in January 2011. This price trend is consistent with other types of recycled paper.

UK exports of recovered fibre have grown rapidly in recent years. The most rapid growth was in exports to Asia, particularly China, which grew from almost zero in 2000 to become the destination of 60 per cent of the UK's recovered fibre export in 2013. The export trend is expected to persist over the years as the prices offered by international mills justify the logistics cost. But the market is getting tougher because it is becoming more selective and this is the nub of the dilemma the world is facing – the balance between the economics of recycling and the pursuit of recycling driven by regulation as an end in itself through our desire to protect the environment and the planet's finite resources.

Governments set annual targets and every community collection programme is driven to achieve those targets. To reinforce them we have increasing landfill charges, but a question we have to ask is: how do these regulations impact on the countryside which may be sparsely populated? It does not become economically friendly for a waste management company to operate on that basis. In London, Paris or New York all the big companies are more than happy to collect but in other less densely populated areas obviously there is much less waste to collect and the economics of collection for a private company are not favourable. Some legislation from Brussels will be economically challenging. The EU Waste Framework Directive says that: from 1 January 2015, waste collection authorities must collect waste paper, metal, plastic and glass separately. It also imposes a duty on waste collection authorities, from that date, when making arrangements for the collection of such waste, to ensure that those arrangements are by way of separate collection.

The reason I mention this very specific example of collections in rural parts is that it is clear to me that different problems, locations and situations surely require different recycling solutions, and regulatory drivers are forcing various regions to promote unrealistic methodologies demanding multiple bins resulting in different collection dates. Where the population is small, producing say 300–400 kilos per person per year, it may make more economic sense for those areas to go to landfill or incineration. The conflict is economics versus environmental ideals.

In France and Sweden for example there is a market for 'environmentally friendly' wood chips for burning. One producer I spoke to said that the wood chips cost more than the coal but people still burn the wood chips because they argue it is more environmentally friendly. There is a danger that we go to great lengths to prove the environmental benefits of some recycling rather than consider the economics. Recycling is necessary but we must not take the restricted view that everything must be recycled without proper consideration of the economics.

There is also the very specific problem for Europe as new member nations join the union, many of which have variable standards of waste collection and recycling tending in large part to rely on landfill alone. The Environmental Services Association (ESA) was quoted as saying that given the disparity in performance among current Member States, it would not be advisable to raise significantly existing targets.[37]

Recycling has got to be talked about alongside climate change. Recycling is such an important element and yet in some parts of the world we haven't even started recycling.

It must make more sense to focus our attention and energies on areas where there is mass consumption and no recycling rather than trying to hit zero waste targets in places like Switzerland where the cost of recovering that last tonne is too expensive to achieve. It is possible to go up to a certain level of recovering recycled waste, but thereafter every additional tonne and every additional percentage is not economically viable and yet we seem determined, even compelled by law, to try. Germany recycles about 74 per cent but for the country to reach 75 per cent would be extremely difficult. It would be easy to slip down to 72 per cent but trying to reach more than 75 per cent would not make economic sense.

The price that we get for a tonne of the recyclable material that we generate has got to take into account, along with the cost of collecting the material, the cost of bringing the material back to the sorting plant, the cost of sorting and baling it and the cost of trucking or shipping it to the ultimate reprocessor which could be in the UK, China or anywhere in Asia. Returning to my example of the sparsely populated area, if volume collected per week is only 10 tonnes it doesn't make sense to insist on elaborate recycling programmes if the financial and environmental expenditure exceeds the small profit margin the waste management company can make.

This is what has been dubbed 'Wastenomics'. We all acknowledge the importance of recycling but it should not be pursued with a missionary zeal which ignores basic economics, recycling for the sake of recycling, producing a product that cannot sustain such expenditure. I repeat the amount of money and energy we expend to achieve that last percentage of recycling can be saved and used to promote recycling in regions of the world where recycling is not even being looked at. To go from 0 to 50 per cent is much easier, 50–75 per cent is more difficult but 75–100 per cent is very difficult. We seem to be committed

---

37   Resource Association 10 September 2013

to a race in Europe to move from 75 to 100 per cent and the cost of moving into that last quartile is increasing rapidly.

There are consequences as recyclers calculate what can and cannot be afforded. For some sectors it is going to mean painful consolidation as Björn Grufman, BIR's President, explains based on his long experience of scrap metal recycling in Northern Europe. He warns some companies will not survive:

> With more than 30 years in the metal industry and mainly in the recycling sector, people may be surprised when I say that I have never before experienced such a dramatic reduction in availability of scrap. The financial crisis and the fast expansion of the product producing industry in the Far East have made the European industry experience a large over capacity. The necessary reduction in industrial production has also naturally reduced the availability of scrap metal.
>
> The lack of material has also resulted in margins being reduced. The obvious result of this reduction in both volumes and margins are of course very negative for the scrap industry. For many of us it is a question of survival.
>
> However, I remain an optimist. I believe the scrap industry in northern Europe has a bright future. As soon we have adapted to this new situation and we have made the necessary reductions which will inevitably mean the merger of several companies we will be able to develop the metal scrap industry with new ambition to serve all our customers, and enjoy a new, brighter future. I have no doubt that our industry has all advantages that will be important in securing that future.

We have experienced an extraordinary growth in waste collection which has been both a blessing in terms of the finite resources saved but a curse in handling capacity. In Europe we were not able to use all this material so it had to be exported outside Europe. Similarly in the UK and elsewhere new regulatory controls resulted in even more waste coming in which in turn led to the expansion of MRFs to create recycled products out of the community collection programme which in the case of paper could be defined as a usable product by a paper mill.

But the world is changing and over the years those paper mills are finding that the cost of production to them, which they quantify in terms of their yield factor – the fibre yield they are getting for the waste they are processing – is poor and the reason for that is quality, which as I have suggested is key. The original quality of the product coming out of MRFs was merely a necessity of handling all the waste that was suddenly being collected throughout the Western economies. Now we have the twin problem of rising waste collections driven by new and more stringent targets combined with the insistence by end users of a higher quality of waste. In the UK alone we are collecting close to ten million tonnes of paper while in Europe the figure is well over 70 million tonnes. All this waste needs to be processed and needs to be used by paper mills but we cannot use all the paper we are recycling so it has to be exported and until 2012/13 we had a ready market in the Far East. That is now changing so the MRFs that we built to handle bulk waste and produce average quality recyclables are no longer up to the standard being demanded by the recipient countries.

The ability to find a home for low grade mixed paper is going to become a challenge. China, India, Indonesia, in fact all the traditional markets, are now very heavily quality focused while we in Europe are doing the opposite by trying to recycle increased volumes of collection with less focus on quality. In the next four to five years the operators of old style MRFs will have a serious problem in marketing their products because on the one hand they are too costly to handle and on the other they are producing a product which the end recipient is not going to buy because it is classified as non-usable waste. Every country is now bringing in a tough regime of quality control. The time is rapidly coming when we must establish an equilibrium between quantity and quality, when we must decide how much we should recycle and decide what percentage should be shifted away from recycling to incineration, landfill or whatever is effective. In fact I would say we have already reached that tipping point.

This has to be reconciled with the increasingly onerous directives and legislation. Countries like America, the UK and other parts of Europe are imposing their own tough export controls to ensure that no one is exporting waste that is not definable as having the right specification and that only material defined as not having reached the end of its valuable life should be recognised as raw material. This alone has caused disagreement in Europe where the EU has endeavoured to bring clarity and harmony. This is important because the lack of agreement raises legal uncertainty where interpretations are not always compatible from state to state when waste is moved from country to country which in turn affects potential investment in new treatments by waste companies. Changes in proposed end-of-waste (EOW) criteria led to protests by the Confederation of European Paper Industries (CEPI), for example, outside the European Commission's Brussels headquarters in September 2013. CEPI feared they would lose control over the quality of the raw material if unrecycled paper and packaging was declared 'End of Waste' and recycled by collectors rather than the paper mills. They said it would inevitably lead to deterioration in quality. Regrettably, the EOW regulations for waste paper are still pending ratification by the Parliament in Brussels.

The impact of this changing climate for MRFs could prove terminal. The ones that are unable to provide 100 per cent good material clear of contamination or those grades to be within allowable range will struggle to survive. When countries like China came in with their Green Fence controls MRFs not able to produce the correct standard could not market their product and they are now being forced either to close or increase investments in more mechanised sorting systems in order to produce a product which would have a market. If they cannot change the quality of what they produce then the waste will end up in incineration or for waste-to-energy purposes. In short it doesn't even need to go through a MRF. China is a market which no recycling nation can ignore – scrap was the top US export to China by value ($11.3 billion) in 2011.[38]

The UK and no doubt other countries as well will need modern MRFs which are able to remove from the sorting stream prohibited and contaminating material which in the past may have been included as acceptable. Plastic, wood, metal cans and clothing items have all got through in the past but now because the tolerance that is allowed is sometimes as low as 1 per cent or even 0 per cent, if it doesn't meet their criteria then the product is not going to be accepted.

This is a challenge for even the biggest operators who have to decide whether investing in new technology to upgrade an old style MRF is worthwhile. A good MRF running at about 10–15 tonnes per hour, working about 16 hours a day is only processing about 200 tonnes a day. Working 7 days a week this is about 1400–1500 tonnes, about 80,000 tonnes per year, which is good enough for a city like Northampton, UK, but the ability of an 80,000 tonne MRF to produce excellent quality recyclables and capture every bit of recyclable waste passing through it is not going to justify the levels of expenditure required to upgrade. Coupled with that, local authorities are tendering for the waste which they award to the highest bidder whereas in the past the waste management companies managed to get most of the collections at very low prices. So there is a cost factor for the baler: first for the money they pay the council and second for the cost of trucking it to the plant, putting the waste through the system and still coming out with a product which is within the tolerance limits. The total cost becomes so high and, with increasing investment costs and global competition, it becomes a challenge to provide the recyclable at the right price. This price itself is capped by the price at which the paper mill can sell the finished paper.

In the long term either MRFs will have to be extremely sophisticated or we will see a gradual demise of the production of the very low end of paper quality coming out of MRFs. Instead of producing it for recycling purposes they will still produce a grade at a cheaper cost but use it for RDF.

38   *The Guardian* 6 September 2013

Where are the drivers, the legislation and all the directives taking us? It is to the inevitable conclusion that all types of recycling, and within that I include incineration as well as landfill sites, will have to be used. It is the obvious conclusion when faced with a market which is ever more demanding, by laws which will only increase waste collection and by end users who are increasingly selective. When I say that recycling makes economic sense it only makes sense when we know we have the raw material, a conveniently located processing plant and above all a ready market for what we produce.

Talk of perpetuating landfills should not cause alarm because they would only be used for material which has passed every process, probably been recycled several times and has come to the end of its useful life and therefore represent a volume which I would expect to be very much lower than we saw 10–15 years ago. But logically it also means that landfill tax should not be so prohibitive that they could never be used. Switzerland has banned all landfill but that can only mean that they are sending their residual waste to someone else or burning it. The challenge is not so much that there is not enough land, although that is clearly true in certain countries such as Singapore or the UAE and Dubai where land is reclaimed from the sea and is very precious, but that the waste we are generating has a value in it and that value should be converted usefully to help the environment in terms of preserving its natural resources and at the same time making the cost of the end product cheaper.

Perhaps the last driver which trumps all others is the global economy. The financial collapse of 2008 changed everything and yet at the same time confirmed what we should have realised – low-quality commodities were no longer in demand. Roy Hathaway of the UK Department for Environment, Food and Rural Affairs, told a conference[39] that the quality of material would play an increasingly pivotal role in trade, with the market set to face short-term financial constraints. He said it was going to be the low-quality end of the spectrum which was going to be squeezed out by an economic downturn.[40] Single-stream collection worked in the beginning: between 1990 and 2003 China experienced a 68 per cent growth in demand for pulp and recycled paper grade, but that dropped to a projected figure of 33 per cent between 2003 and 2010.[41] China no longer has the unlimited supply of cheap labour to sort the contaminated waste and they are also producing their own waste and virgin products which they need to handle. China collected about 20 million tonnes of paper for recycling in 2011 – a big increase on the

39 WRAP Conference, 23 October 2008, Issue: Focusing on navigating the current economic downturn.
40 Container Recycling Institute Report: *Understanding economic and environmental impacts of single stream collection systems*
41 Container Recycling Institute Report: *Understanding economic and environmental impacts of single stream collection systems*

previous year.[42] Consequently fibre export into China will fall as the snapshot of the European paper trade with China shows (see Table 2).

Table 2 The changing trends of European fibre exports to China – a six month comparative snapshot

EU Country	2012	2013	
UK	1,905,771	1,579,675	DECREASE
Holland	1,172,998	797,581	DECREASE
Italy	622,272	477,270	DECREASE
Belgium	430,834	341,996	DECREASE
France	309,045	253,669	DECREASE
Spain	292,099	216,197	DECREASE
Germany	309,045	253,669	DECREASE
Ireland	90,389	102,805	INCREASE
Greece	56,266	50,118	DECREASE
Portugal	40,762	29,934	DECREASE
Norway	32,493	25,911	DECREASE
Turkey	11,415	4,687	DECREASE
Sweden	9,189	5,478	DECREASE
Poland	14,429	4,486	DECREASE
Slovenia	8,976	12,859	INCREASE
Bulgaria	2,942	3,613	INCREASE

We have to accept that the world has changed and that we all operate on a single planet. Rules and regulations have to make both economic as well as environmental sense and no one can operate in isolation pursuing their own idealistic path. It is not a crisis if we don't want to make it one but we have to use all the tools at our disposal to handle an unlimited raw material which can both help us as well as cause great harm through its misuse. We must not tie ourselves down with impossibly rigid legislation but we must recognise that waste is a constantly changing commodity which can be harnessed by innovative technology. It has almost become a force of nature, certainly a force created by human nature, and it is both volatile and valuable. Handle with care.

---

42   BIR – Recovered Paper Market in 2011

5

Waste stream

If quality is now the watchword, what impact will that pursuit of perfection have on the waste stream and the recycling industry as a whole? First of all we should understand what we mean by the waste stream, just two simple words describing a surprisingly complex 'raw material'.

The US Environmental Protection Agency provides a succinct description: 'Waste stream – the total flow of solid waste from homes, businesses, institutions, and manufacturing plants that is recycled, burned, or disposed of in landfills.'

Contained in that short sentence is a vast mix of waste which is divided and sub-divided into different categories. Table 3, based on statistics from the UK's Department of Environment, Food and Rural Affairs (DEFRA), gives a snapshot of waste materials between 2004 and 2008 – each one of these categories has its own subset of grades. In 2008, a total of 289 million tonnes of waste was generated in the UK from various sources including construction (35 per cent), mining (30 per cent), industrial and commercial (23 per cent), household (11 per cent) and secondary sewage waste (1 per cent).

That is the range of material that every city and municipality around the world typically has to deal with and it is safe to say that the majority are not coping, having to rely on passing much of their waste to somebody else, but the waste that is exported is beyond the control of the exporting countries. Although by law it is supposed to be carefully recycled, DEFRA has admitted that most of the waste shipped out of the UK regardless of how diligently it may be sorted is still heading for landfill.[43]

Let us consider some of the major waste categories.

Paper

As we have noted paper is by far the biggest component of waste collection. The main producers and consumers of paper and board are China, USA, Japan and Germany. But every nation in the world needs paper and consequently we all

---

43   *Daily Mail* 6 April 2013

Table 3 Types of waste collected from waste streams

EWC-Stat Ver 3 Code	Waste type (h=hazardous)	2004	2006	2008
01,02,03	Chemical wastes	2,539	1,999	1,419
01,02,03 excl 01.3	Chemical wastes excluding used oils (h)	3,158	2,407	1,821
1.3	Used oils (h)	463	396	399
5	Health care and boilogical wastes	174	201	217
5	Health care and biological wastes (h)	269	356	350
6	Metallic wastes	8,150	6,649	4,433
6	Metallic wastes (h)	32	35	34
7.1	Glass wastes	2,119	2,356	2,317
7.1	Glass wastes (h)	6	-	0
7.2	Paper and cardboard wastes	12,524	13,727	12,298
7.3	Rubber wastes	176	212	286
7.4	Plastic wastes	2,039	3,373	2,371
7.5	Wood wastes	3,980	6,943	3,650
7.5	Wood wastes (h)	14	11	22
7.6	Textile wastes	377	243	269
7.7	Waste containing PCB (h)	0	0	4
0.9 exc. 09.11, 09.3	Animal and vegetal wastes	8,105	9,377	9,665
9.11	Animal waste of food preperation and products	1,665	1,836	2,254
9.3	Animal faeces, urine and manure	118	115	83
10.1	Household and similar wastes	52,363	46,364	42,537
10.2	Mixed and undifferentiated materials	2,686	5,729	2,603
10.2	Mixed and undifferentiated materials (h)	191	458	129
10.3	Sorting residues	674	581	2,452
10.3	Sorting residues (h)	1	-	6
11	Common sludges	29,731	17,037	18,133
12	Mineral wastes	189,358	182,407	176,565
12	Mineral wastes (h)	646	856	1,586
8.13	Other wastes	3,497	277	678
8.13	Other wastes (h)	222	3,101	1,974
	Total, non-hazardous	**320,274**	**299,424**	**282,230**
	Total, hazardous	**5,003**	**7,629**	**6,325**
	Total, general	**325,277**	**307,053**	**288,555**

*Indicative figures

have to dispose of it one way or another and because there is such a wide choice only the very best quality is wanted. Not content with its Green Fence crackdown on contaminated shipments, China went one step further in October 2013 by launching what it called its 'Earth Goddess – Phase Three' programme to combat the illegal smuggling of hazardous waste from Europe and North America to the Asia-Pacific region. The Xinhua news agency reported that the Green Fence clampdown had resulted in the seizure of '33,500 tonnes of waste batteries, waste slag, waste paint, tyres, old clothes and other foreign garbage'. They will find more.

The clampdown was inevitable. Speaking at the Paper Recycling Conference in Chicago in October 2013, Bill Caesar of Waste Management Inc., Houston said: 'The nature of materials [Chinese buyers] were dealing with, some of it was less than desirable...things had to change.'[44]

---

44 *Recycling Today*, 16 October 2013

But we should not be surprised. The Green Fence is not a new concept. In 2007/8 China already had their own specification of the grades they wanted. At the same time Europe and America also had their own specifications. The problem is that when we export from these regions we have been too focused on our own requirements, in effect trying impose our standards on others. But China now have their own standards and because their own domestic industry has built up to such a level – they are collecting around 50 million tonnes of local waste paper – they are correctly demanding that the quality of imported fibre must meet their quality standards. It makes complete sense only to import fibre that is as good or better than domestic quality. So the Green Fence is merely reinforcing their standards which they have been striving for long before the new policy came into force. Now India has come out with its own controls defining the levels of non-permissible material within the recyclable grade. It doesn't matter how anyone else interprets their waste, if it does not meet their standard it will not cross their borders. All these traditional recipient countries are introducing some sort of quality control and we must make adjustments if we want to stay in the industry. It is an initiative which should be universally welcomed as it weeds out the illegal operators.

Table 4 India: permissible limits of non-recyclables

Sr. No.	Grade	Limit (percentage) prohibitive contents
1	Residential mixed paper	2
2	Soft mixed paper	1
3	Hard mixed paper	½
4	Boxboard cuttings	½
5	Mill wrappers	01-Feb
6	News	1
7	New, de-ink quality	None permitted
8	Special news, de-ink quality	None permitted
9	Over-issue news	None permitted
10	Magazines	1
11	Corrugated containers	1
12	Double sorted corrugated	½
13	New double-lined kraft corrugated cuttings	None permitted
14	Fibre cores	1
15	Used brawn kraft	None permitted
16	Mixed kraft cuttings	None permitted
17	Carrier stock	None permitted
18	New coloured kraft	None permitted
19	Grocery bag scrap	None permitted
20	Kraft multi-wall bag scrap	None permitted
21	New brown kraft envelope cuttings	None permitted
22	Mixed ground wood shavings	None permitted
23	Telephone directories	None permitted
24	White blank news	None permitted
25	Ground wood computer printout	None permitted
27	Flyleaf shavings	None permitted
28	Coated soft white	None permitted
29	Hard white shavings	None permitted

Sr. No.	Grade	Limit (percentage) prohibitive contents
30	Hard white envelope cuttings	None permitted
31	New coloured envelope cutting	None permitted
32	Semi bleached cuttings	None permitted
33	Unsorted office paper	2
34	Sorted office paper	1
35	Manifold coloured ledger	½
36	Sorted white ledger	½
37	Manifold white ledger	½
38	Computer printout	None permitted
39	Coated book stock	None permitted
40	Coated ground wood section	None permitted
41	Printed bleached board cuttings	½
42	Misprinted bleached board	1
43	Unprinted bleached board	None permitted
44	Bleached cup stock	None permitted
45	Printed bleached cup stock	None permitted
46	Unprinted bleached plate stock	None permitted
47	Printed bleached plate stock kinds	None permitted
48	Speciality grades (white waxed cup cuttings, plastic coated cups, printed waxed cup cuttings, polycoated bleached kraft-unprinted, polycoated bleached kraft-printed, polycoated milk carton stock, polycoated diaper stock, polycoated box board cuttings, waxed boxboard cuttings, printed and/or unprinted bleached sulphate containing foil, waxed corrugated cuttings, wet strength corrugated cuttings, asphalt laminated corrugated cuttings, beer carton scrap, contaminated bag scrap, insoluble glued free sheet paper and/or board, whate wet strength scrap, brown wet strength scrap, printed and/or coloured wet strength scrap, file stock, new computer printout, ruled white, flyleaf shaving containing hot melt glue, carbon mix, book with covers, unsorted tabulating cards, carbonless treated ledger, plastic windowed envelopes, textile boxes, printed TMP, unprinted TMP, manilla tabulating cards, sorted coloured ledgers)	None permitted

Waste paper or fibre is defined by quality, and quality we define by various factors. Some of the major factors we consider are: moisture content, contamination and outthrows (low grade material which in large quantities become contaminants). There are two main categories of waste paper, pre-consumer and post-consumer, and these can be further divided into some 200 different grades. Just in case you thought waste paper was only the difference between white, brown and cardboard here is an alphabetical list which is not exhaustive as there are other speciality grades:

Aseptic Packaging	Box Board with Foil
Beer Carton Waste	Box Board with Poly
Bleached Coffee Filter	Brown Kraft
Bleached Cup Stock (#1 Cup)	Carbon Interleaved Ledger
Bleached Kraft	Carbonless Thermal Ledger Paper
Bleached PE-Coated Board (European Grade)	Carbonless Copy Paper (European Grade)
	Carrier Stock (Grade #17)
Book & Book Stock	Cartonboard Cutting (IMW) South African
Box Board	Grade
Box Board Cutting (Grade #4) ISRI Grade	Coated Book Stock (CBS) (Grade #43)

Coated Flyleaf Shavings (Grade #27)
Coated Groundwood Sections (CGS)
Coated Soft White Shaving (SWS) (Grade #28)
Colored Envelope
Coated Groundwood Sections (CGS) (Grade #44)
Colored Kraft
Colored Ledger
Colored Letters (European Grade)
Colored Tab Cards
Colored Woodfree Magazines (European Grade)
Computer Print Out (CPO) & Laser Printed White Ledger (Grade # 42)
Corrugated Container (OCC) (Grade #11)
Deinking Grade Newspaper
Double Lined Kraft (DLK)
Double Sorted Corrugated Container (DS OCC) (Grade #12)
Fiber Cores (Grade #14)
Glassine
Grey Board (European Grade)
Grocery Bag Scrap (KGB) (Grade #19)
Groundwood Computer Printouts (GW CPO) (Grade #25)
Groundwood Fiber
Hard Mixed Paper (HMP) – ISRI Grade
Hard White Envelope Cutting (HWEC) (Grade #31)
Hard White Shaving (HWS) (Grade #30)
High Grade White One (W1) South African Grade
Heavily Printed White Shaving without Glue (European Grade)
Heavy Letter One (HL1) South African Grade
Heavy Letter Two (HL2) South African Grade
High Grade White One (W1) South African Grade
Kraft Board Stock
Kraft Envelope
Kraft Grade Corrugated Containers (K4) South African Grade
Kraft Grocery Bags (KGB) (Grade #19)
Kraft Multi-wall Bag Waste (Grade #20)
Kraft Multi-wall Poly Bag Waste
Lightly Printed Bleached Kraft (LPBK)

Lightly Printed Bleached Sulphate Board (European Grade)
Lightly Printed White Shaving (European Grade)
Lightly Printed White Shaving without Glue (European Grade)
Liquid Board Packaging (European Grade)
Magazines (SBM) South African Grade
Magazines (OMG) Grade # 10
Magazines with Hot Melt
Manila File Folder Stock
Manila Tab Cards
Manifold Colored Ledger (MCL) (Grade #39)
Manifold White Ledger (MWL) (Grade #41)
Mechanical Grades Special News (SN) South African Grade
Mechanical Pulp-Based Computer Print-Out (European Grade)
Mill Wrappers ISRI Grade
Misprinted Bleached Board (Grade #46)
Mixed Envelope (New)
Mixed Flyleaf Shaving (Grade #22)
Mixed Groundwood Shaving (Grade #22)
Mixed Kraft Cutting (Grade #16)
Mixed Lightly Colored Printer Shaving (European Grade)
Mixed Lightly Colored Woodfree Printer Shaving (European Grade)
Mixed Newspaper and Magazines (European Grade)
Mixed Newspaper and Magazines 1 (European Grade)
Mixed Newspaper and Magazines 2 (European Grade)
Mixed Office Paper
Mixed Packaging (European Grade)
Mixed Paper (Grade #2)
Mixed Paper (CMW) – South African Grade
Mixed paper and board, unsorted, but unusable materials removed (European Grade)
Mixed paper and board-sorted (European Grade)
Mixed Recovered Paper and Board (European Grade)
Mixed Tab Cards
Mixed Waste Paper
Multi Printing (European Grade)

Municipal Waste Paper
New Brown Kraft Envelope Cutting (Grade #21)
New Carrier Kraft (European Grade)
New Colored Envelope Cutting (Grade #33)
New Colored Kraft (Grade #18)
New Corrugated Kraft Waste (K3) South African Grade
News (Grade #6)
News Deink Quality (Grade #7 ONP)
New Double Lined Kraft Corrugated Cutting -Dark Medium Light Wax
New Double Lined Kraft Corrugated Cutting (DLK) (Grade # 13)
New Kraft (European Grade)
New Kraft Multi-Wall Bag (Grade # 20)
New Shaving of Corrugated Board (European Grade)
Newspapers (European Grade)
Old Corrugated Cardboard
Old Corrugated Container (OCC) (Grade #11)
Old Corrugated Container (European Grade)
Old Magazine (OMG)
Old Newspaper – ISRI Grades
Other Waste Paper
Other PE-Coated Board (European Grade)
Over Issue Newspaper OI or OIN (Grade #09)
Over Issue News (#9 ONP) – ISRI Grades
Over Issue News (FN) South African Grade
Plastic Windowed Colored Envelope
Plastic Windowed Kraft Envelope
Plastic Windowed Printed Kraft Envelope
Plastic Windowed White Envelope
Post-Consumer Waste Paper
Pre-Consumer Waste Paper
Printed Bleached Board Cutting (Grade #45)
Printed Bleached Cup Stock (#2 Cup)
Printed Bleached Kraft (PBK)
Printed Bleached Plate Stock (Grade #51)
Printed Bleached Sulphate Board (European Grade)
Printed Kraft Envelope
Printed White Wet-Strength Woodfree Papers (European Grade)
Publication Blanks (CPB) (Grade #26)

Recovered Paper
Recovered Paper Grades
Regular News, De-ink Quality (#7 ONP) – ISRI Grades
Residential Mixed Paper – ISRI Grade
Ruled White
Scrap Paper
Semi Bleached Cutting (Grade #35)
Shredded Mixed Office Paper
Soft Mixed Paper (Grade #1)
Soft Mixed Paper- ISRI Grade
Soft White
Solid Fiber Containers
Sorted Graphic Paper for Deinking (European Grade)
Sorted Office Paper (European Grade)
Sorted Office Paper (SOP) South African Grade
Sorted Office Waste (SOW)
Sorted White Ledger (SWL) (Grade #40)
Special Grade Liquid Board Packaging (LBP) South African Grade
Special Magazine (SSBM) South African Grade
Special News Deink Quality (#8 ONP)
Special News, De-ink Quality (#8 ONP) – ISRI Grades
Supermarket Corrugated paper and Board (European Grade)
Supermix (SMW) South African Grade
Tear White Shaving (European Grade)
Telephone Books (European Grade)
Telephone Directories (TD) South African Grade
Telephone Directories (Grade #23)
Unprinted Bleached Board (Grade #47)
Unprinted Bleached Plate Stock (Grade #50)
Unprinted Bleached Sulphate Board (European Grade)
Unprinted Bleached Sulphate Kraft
Unprinted White Wet-Strength Woodfree Papers (European Grade)
Unsold Magazine (European Grade)
Unsold Magazine without glue (European Grade)
Unsold Newspapers (European Grade)
Unsold Newspapers, no Flexo-Graphic Printing allowed (European Grade)
Unsorted Office Paper (UOP) Grade # 37

Unused Corrugated Kraft (European Grade)	White Ledger
Unused Corrugated Material (European Grade)	White Lightly Printed Multiply Board (European Grade)
Unused Kraft Bags (K1) South African Grade	White News Blank (WNB) (Grade #24)
Unused Kraft Sacks (European Grade)	White Unprinted Multi-ply Board (European Grade)
Unused Kraft Sacks with Polycoated Papers (European Grade)	White Woodfree Books (European Grade)
Used Brown Kraft (Grade #15)	White Business Forms (European Grade)
Used Corrugated Kraft 1 (European Grade)	White Business Forms (European Grade)
Used Corrugated Kraft 2 (European Grade)	White Mechanical Pulp Based Coated and Uncoated Paper (European Grade)
Used Kraft (European Grade)	White Mechanical Pulp Based Paper containing Coated Paper (European Grade)
Used Kraft Bags (K1)	White Newsprint (European Grade)
Used Kraft Sacks (European Grade)	White Newsprint (European Grade)
Used Kraft Sacks with Polycoated Papers (European Grade)	White Shaving (European Grade)
Waxed Box Board	White Woodfree Computer Printout (European Grade)
Waxed Corrugated	White Woodfree Shaving (European Grade)
Waxed Cup Stock	White Woodfree Uncoated Shaving (European Grade)
Wet Labels (European Grade)	Woodfree Binders (European Grade)
White Blank News (WBN) (Grade #24)	Wrapper Kraft (European Grade)
White Envelope	
White Heavily Printed Multiply Board (European Grade)	

Not only are the recovered fibre grades multifarious but pulp is also produced in various grades. Here is another list of various pulp grades:

Ground Wood Pulp	Viscose Pulp
Refiner Mechanical Pulp	Alpha cellulose Pulp
Chemo Mechanical Pulp	Rayon grade Pulp
Chemo Thermo Mechanical Pulp	Hardwood Pulp
Stone Ground Wood Pulp	Soft wood Pulp
Bleached Pulp	Semi-bleached Pulp
Unbleached Pulp	NSSC Pulp
Kraft Pulp	NBSK Pulp
Sulfite Pulp	Northern Bleached Softwood Kraft Pulp
Sulfate Pulp	Northern Bleached Hardwood Kraft Pulp
Wood Pulp	Southern Bleached Softwood Kraft Pulp
Non-wood Pulp	Southern Bleached Hardwood Kraft Pulp[45]
Fluff, dissolving grade Pulp	

45   www.paperontheweb.com

Little wonder then that paper dominates the waste stream.

Although in waste paper we are dealing with more volume, the challenges of contamination are quite difficult so it is important that the producers of waste and waste contractors understand that they have a duty of care to ensure that methods used in waste collecting do not become a liability further down the sorting chain making it more difficult or more expensive to segregate. This duty of care or responsibility is an area we will consider further.

The question for the paper recycler is what impact will the increasing demand for high quality have on the industry? Will some grades simply have to be dropped?

Within the spectrum of waste paper the high grades are the pure white cuttings which come from the printers and at the other very low end is the paper from households which is more contaminated. The end users will only take the low end provided the plastic content is less than 0.5 per cent – some countries like India are now saying it should be zero per cent and free of plastic coated material. This is a high threshold. Today some of the cheaper end paper has a plastic coating to give more tension and strength but plastic cannot be melted down during the processing of the paper into fibre in the paper mill. A bale of paper might also include contamination such as clothes or pieces of metal. All this foreign material takes time to sort and also more energy in terms of machine time to sort. And so now all the recipient countries who increasingly will also have their own collections of similar low grades of paper which they are already struggling to deal with are refusing to accept such waste. The logical consequence of producing more than can be consumed domestically and a failure to find an export market means that the waste will end up in landfill.

This is why I say there must be a shift in priorities. The challenge is now moving more upstream than downstream where the responsibility should be on the collector of waste or on the industry or householder that is producing the waste so they become more connected with, and have a better understanding of, the product that the waste is going to be used for, in this instance by a paper mill. It may sound harsh but to a large extent the producer of waste has no understanding of where the waste is going and how it connects in the production cycle, the paper mill or the processing centre.

Until now mixing different types of waste paper has been acceptable – not any longer. At its highest level recyclables will have white paper such as envelopes separated from cardboard, and newspaper should be separated from brown paper and white paper because white paper is a purer form of paper which is referred to as a pulp substitute. It not only sells at a higher price, it also costs more to produce. Naturally in a household one would not have so many different items of paper waste and segregation would have to take place at a recycling plant where they have the ability to sort albeit mechanically and manually because not every form of paper can be separated automatically. A modern, state-of-the-art facility will soon become a necessity for all.

Plastic

By contrast plastic sounds simpler in that before recycling it is pre-sorted by polymer types which were introduced by the Society of Plastics Industry in 1988. The seven main groups in which plastic can be classified are:

1. Polyethylene terephthalate (PET) – commonly used for soft drink bottles.
2. High density polyethylene (HDPE) – used most often for milk and water containers because of its resistance to moisture; it is also used for shopping bags.
3. Vinyl (polyvinyl chloride or PVC) – highly versatile. It is found in rigid products in the construction industry e.g. pipes as well as cling film.
4. Low density polyethylene (LDPE) used for film applications, flexible and commonly used for container lids.
5. Polypropylene (PP), which has a high melting point, is used in products as varied as packaging, thermal underwear, carpets and polymer banknotes.
6. Polystyrene is typically used in protective packaging, containers, cups and lids.
7. Other. This category as its name implies is used for products made from any other resin not included above.

Symbol	Acronym	Full name and uses
1	PET	Polyethylene terephthalate - Fizzy drink bottles and frozen ready meal packages.
2	HDPE	High-density polyethylene - Milk and washing-up liquid bottles.
3	PVC	Polyvinyl chloride - Food trays, cling film, bottles for shampoo, mineral water and shampoo.
4	LDPE	Low density polyethylene - carrier bags and bin liners.
5	PP	Polypropylene - Margarine tubs, microwaveable meal trays.
6	PS	Polystyrene - Yoghurt pots, foam meat or fish trays, hamburger boxes and egg cartons, vending cups, plastic cutlery, protective packaging for electronic goods and toys.
7	Other	Any other plastics that do not fall into any of the above categories. For example melamine, often used in plastic plates and cups.

They are all identified by their initials and corresponding number – their Plastic Identification Code (PIC) – in the classic recycling symbol of the three chasing arrows, for example HDPE 2. But while it may be easy to classify it is no easier to handle once it becomes part of the waste stream.

In September 2012 my colleague on the Bureau of International Recycling and Chairman of our Plastics Committee, Surendra Borad, warned:

> In 2010 Europe produced about 24 million tonnes of scrap plastic. Only about 6 million tonnes of this is recycled, half in Europe and the other half elsewhere. As Europe does not have sufficient processing capacity, we are left with a lot of plastic for export. If other countries close their borders, for example in retaliation for our export restrictions, then we will be left with a gigantic mountain of surplus plastic.

This is clearly going to be a sticking point for the entire waste stream in the years to come. As countries become more demanding, as they develop their own recycling industries and prices become more competitive internationally, I fear there will be a tendency to become more nationalistic, even protectionist. This will not help anyone and it will not help achieve stable prices as the whole world struggles to preserve finite resources. Any restriction on the exports of plastic scrap from Europe may lead to lower price realisation and this in turn will lead to lower collection of scrap. As a result the European recycling target may not be achieved.

Dominique Maguin, former President of the BIR, pointed out what happens when politics interferes with the industrial process:

> (If) the price is fixed arbitrarily, by consideration of protectionist policy or national preference, to promote such business, even as an industrial player, then these recycling activities will be doomed to failure as unprofitable. The establishment by political decision without any preparation or exchange with professionals involved in the widespread collection in Germany in the late 80s and then in France and in other economies a few years later, led to an historic collapse of commodity prices of the recycled paper-cardboard industry.

Plastic is more and more in demand and is being used in innovative ways; however the increasingly exotic waste streams into which this plastic goes – cars, furniture and clothing grade polyester – are not linked with the recycling process so we are getting more mixed grades of plastic which used to go to Asia to be sorted manually. It was very worthwhile as there is such a difference in price between a shrink wrapper and a milk bottle and between a clear milk bottle and a soft drink bottle. Understandably to most people it is all just plastic and is mixed together. But the challenge in the plastic industry is exactly the same as the waste paper industry in that we have to find ways of sorting these different streams into more usable and identifiable grades; contamination includes the sugary liquid in many soft drinks bottles which have to be prewashed, a facility not all recycling plants possess. The technology to sort by colour and type exists but the operator has to decide how many if any optical colour sorters to put along a conveyor belt and that will depend on the volume handled before the investment can be justified.

Plastic is one of the most expensive and complicated wastes to recycle. In a dedicated facility the process of recycling plastic is first sorting and washing, then shredding, classifying and extruding. Sorting is important because different plastics mixed together produce a poor quality product. The extrusion element involves melting the sorted and classified waste into pellets which are sold to manufacturers. A good state-of-the-art facility may cost anything between $10 and 20 million which may be one reason why we seem to be so reluctant to build more, although plastic clearly has a value: worldwide trade of recyclable plastics is worth $5 billion per year and is estimated to represent a total of 12 million tonnes.[46] But Ton Emans, President of the Plastics Recyclers Europe, warned that current recycling rates were no longer acceptable. 'In the long run, economic growth, demographic changes and growing scarcity of raw materials will not allow Europe the luxury of wasting 76 per cent of all plastic materials used.'[47] That is a staggering amount and sadly it does not all end up in landfills.

It is estimated that 100 million tonnes of waste is floating in the Pacific Ocean covering an area twice the size of the continental United States and making the ocean the world's biggest 'landfill site'. This 'plastic soup' off the California Coast across the northern Pacific, past Hawaii and almost as far as Japan is held in place by underwater currents in the middle of what is called the Subtropical Gyre. Once plastic is sucked into this vortex of currents it can never escape and simply breaks down into minute plastic particles. Fourteen billion lbs of garbage, mostly plastic, is dumped into the oceans every year.[48]

Plastic's value as a recycled product is not in question. One tonne of recycled plastic saves 5,774 kWh of energy, 16.3 barrels (2,604 litres) of oil, 98 million Btus of energy, and 22 cubic metres of landfill.[49] And the plastics industry has a good reputation for doing its bit for the planet. According to the British Plastics Federation, between 1991 and 2000 the average weight of plastics film (g/m²)

46   Bureau of International Recycling (BIR)
47   *European Plastics News* 13 September 2013
48   *Earth in Danger: Pollution by* Helen Orme, Bearport Publishing New York, NY 2008
49   BIR

decreased by 36 per cent whilst the average weight of bottles and containers decreased by 21 per cent.

Metal

Other waste streams have their own unique issues but the difficult equation of volume, quality and profit is never far away. The metal recycling industry has always been one of the most important with an estimated turnover in the UK alone of £5.6 billion per annum. In America despite a dip post-2008 the average annual industry turnover between 2013 and 2018 is forecast to be $15 billion.[50]

As Christian Rubach, President of the BIR Ferrous Division, notes, scrap metal is both needed and environmentally friendly:

> Today, 55 per cent of the EU steel production comes from steel scrap, 70 per cent of the US steel production and 88 per cent of the Turkish production. Why is that so? Steel scrap is endlessly recyclable. Steel scrap is emission friendly. Steel scrap is the raw material source of developed economies, which never ends, as long as consumption patterns continue. Steel scrap is the most important raw material source in developed countries, because it is in those historical infrastructure steel based investments like railway systems, buildings, bridges, roads, dams, airports, ships and car bodies which one day will be scrapped to continue the circle.

Scrap metal primarily comes in two forms, ferrous and non-ferrous. The non-ferrous, such as aluminium and copper, is used most commonly for making drink cans, copper wires and tubes, and ferrous for steel and iron. The major non-ferrous metals – including copper and aluminium – have seen an increase in demand from 12 million tonnes and 20.8 million tonnes in 1995 to 19.3 million tonnes and 44.9 million tonnes in 2011, respectively.[51]

Half of all steel manufactured in the world comes from recycled material but although the demand is not questioned – world crude steel production increased in 2012 by 1.2 per cent to 1.55 billion tonnes[52] – the key for recycling companies is the volume that can be collected in any given area. Even the biggest companies have been forced to accept that the economics of metal recycling do not make sense everywhere and have closed their operations.

Worries about protectionism also plague the metal sector. At the World Recycling Convention in Shanghai in May 2013, Zain Nathani of the Nathani Group of Companies warned of the 'surprising' decision by India to increase

---

50   *Euromonitor International* 5 March 2013
51   BIR
52   Worldsteel

basic customs duty on iron and steel scrap from 0 per cent to 2.5 per cent. It was a 'regressive' decision he said which had slowed a once strong flow of ferrous scrap imports. Nevertheless some 570 million tonnes of steel scrap was used globally in 2012. That year Turkey continued to be the world's foremost importer of steel scrap as China limited its imports in line with the country's 'green' agenda and desire to increase its domestic usage.

Some sought to protect their resources by imposing export bans as the CRU Group reported:

> The 2009 Indonesian unprocessed minerals export ban came into force on 12th January 2014, but not all commodities are treated equally. The ban was not implemented as originally proposed. On 8th January, in a partly anticipated turn of events, Indonesia's mining ministry sought to pass legislation – which allowed for continued exports of a selection of unprocessed materials. While Indonesia has previously demonstrated that it has the will to successfully implement development policy to guide domestic industry up the value chain, whether they can hold their nerve this time remains to be seen.[53]

Robert Voss of Voss International and Chairman of BIR's International Trade Council has put the metals sector into historical perspective and highlighted some threats ahead:

> Non Ferrous Recycling is one of the world's oldest industries. Anthropologists speak of the Copper Age and the Bronze Age as stages through which Man has passed on his journey to becoming truly civilised. As soon as metals began to be extracted from the earth and melted, products capable of being recycled appeared and thus recycling of non ferrous metals – primarily copper and bronze – -had begun.
>
> So why is recycling being touted as a 'new industry'? Why do politicians and bureaucrats want to jump onto the recycling band wagon? Why has the world seemingly become 'green' when copper and its alloys along with lead and other non ferrous metals have been recycled for well over 10,000 years?
>
> Controlling the recycling of – or more realistically the movement of – recycled non ferrous metals, often under the guise of environmental controls, has become a political football in recent years as countries have tried to protect what is now recognised by many as a vital raw material of high worth and strategic value…nothing new to us recyclers of non ferrous metals who have been at it for years.
>
> Today the non ferrous recycling industry has come a long way and is a sophisticated industry based on a pyramid structure where scrap still has

53   Peter Ghilchik, CRU Insight January 2014

to be collected and sorted at the bottom of the pyramid whilst the top of the pyramid has become a highly mechanised and structured industry.

Non ferrous scrap is truly a worldwide commodity; every country in the world produces non ferrous scrap, every country in the world can use non ferrous scrap and the price is truly international, usually based on the London Metal Exchange (LME). And most importantly today everyone on the planet who has access to the internet can access a price for all grades of non ferrous scrap. Gone are the days of making money out of people's ignorance – today thanks to technology no one is ignorant.

So the international trade in non ferrous metals has developed to become a service industry where name and reputation, reliability and relationships are as important if not more so than price.

However there is a growing trend towards protectionism for economic reasons (but often under the misrepresented guise of environmentalism) by countries all over the world from Europe to Asia and Africa to The Americas. And the interruption to the free flow of this vital secondary raw material will have a dramatic impact on world trade.

In current times there is a certain lack of available scrap as a result of the recent worldwide recession: people do not change their car, TV or refrigerator and utilities are forced to curtail replacement of cables and pipelines and buildings are not demolished to make way for new ones when times are bad. Furthermore, production scrap is not entering the market in the same volume during, or as we exit a recession. So that is the current situation combined with a lack of future demand as confidence in world economies takes a long time to re-establish. Non ferrous metals are bought in advance during boom times and are often describes as 'The Pendulum of The Economy'…we swing and economies follow…but generally it is the lack of confidence in world economic growth that has the most dramatic effect on supply and demand for non ferrous metals. LME prices (and thus the basis for scrap prices) are no longer solely a reflection of the principals of economics as they once were. Outside influences like hedge funds and speculators can and do have a significant effect on daily price movements which in turn affect the flow of non ferrous scrap.

How does the future look for the secondary non ferrous metals industry in the next decade? There is no doubt that metals will continue to be in strong demand worldwide, in fact there is every indication that demand will increase as emerging economies continue their development. And as the 'mine above the ground' and with more emphasis on green economies, recycling will be around for at least the next 10,000 years!

We have seen a consolidation of stakeholders in the industry over the last decade and I am not sure this will continue. I foresee some of the large conglomerates deconstructing and the growth once again of smaller companies in the supply chain, but they will be efficient companies who know the legislation and have the relationships with the consumers to supply their ever exacting requirements…once again the emphasis being on service.

'There will be new emerging markets for non ferrous metals in the coming decade especially in South America which may affect the amount of scrap flowing to Asia from USA but also much will depend on the state of world currencies in the coming years.

After the recent Euro debacle and the saving of the European currency will it strengthen against the Dollar or will the Renminbi become a dominant player in world currency markets? So macro economic factors will have a significant influence on the world flows of non ferrous scrap.

Legislation will continue to play a key role in the trade as more countries try to protect their secondary raw materials but often without taking account of negative counter measures along with genuine environmental controls – many of which are not designed or even relevant to the non ferrous industry.

The trade in secondary non ferrous metals will become more specialised and more technical in years to come with closer relationships between traders and consumers on whom they will rely for supply. Let us hope that the vital role that this industry plays in world trade is truly recognised and secondary non ferrous metals are allowed to flow freely around the world in one of the truly international marketplaces.[54]

Glass

Glass has its own particular set of issues in part because once a bottle is shattered people no longer seem to treat it as a separate waste item. In single-stream collections probably only about 40 per cent of glass survives the process to be recycled, barely 20 per cent is recovered for low end use and the rest ends up in landfill. As we have seen glass mixed in with paper or plastics causes inevitable contamination and a loss of quality possibly even rejection of the waste. Glass – the different colours are recycled separately – is surprisingly heavy and because of its density makes up a major volume of domestic and industrial waste. Cullet, recycled broken waste glass, makes its own contribution to saving the planet's finite resources: every metric tonne recycled saves 315 kilograms of carbon dioxide being released into the atmosphere.[55]

Textiles

Olaf Rintsch of Textil-Recycling K. & A.Wenkhaus GmbH, Hamburg gives this snapshot of the global textile industry and stresses the need for manufacturers to grasp the importance of reusing material.

---

54   Robert Voss, personal communication
55   Waste Online

Overall it looks quite promising for textile recycling albeit in a falling market. Unrest and wars have led to reduced flow of goods. The warehouses are well stocked, the cash flow is delayed and worldwide falling prices are to be expected.

The main challenge should be to keep the access to original clothing. Legislative regulations such as some in Germany are counterproductive where the legislature attempts to get access to all the collections of used textiles. For Europe it is very important that a standardized legislation is found in the near future. Different execution in the member states leads to isolated applications and confusing market prices.

For our colleagues in the U.S. it is very important to achieve a greater acceptance of textile recycling in their own country. There is much to do to make the importance of this industry better known in the population.

The Indian market should be continuously closely monitored. Especially the working conditions in this country should be fundamentally improved.

For the future it is very important to explain to the producers of new goods the immense importance of using recycled materials. Unfortunately the acceptance is not as satisfying as it would be desirable.

The future will show us that is indispensable to use only 1/2 litre of water for recycling 1 kg of cotton instead of using 27,000 litres of water for 1 kg of new produced cotton. These savings must be achieved to save resources for our future generations.

E-waste

There are many other waste streams, not least toxic waste, which need special treatments, medical and chemical waste and hazardous waste from industry. But probably the fastest growing 'modern' waste is e-waste with its own toxic element.

We tend to think of e-waste as being limited to high-tech items but the category is much broader. The European Union Directive on Waste Electrical and Electronic Equipment –WEEE Directive – has ten categories:

1. Large household appliances (e.g. refrigerators)
2. Small household appliances (e.g. coffee machines)
3. IT and telecommunications equipment (e.g. computers and phones)
4. Consumer equipment (e.g. radio and television sets)
5. Lighting equipment (e.g. fluorescent lamps)
6. Electrical and electronic tools with the exception of large-scale stationary industrial tools (e.g. drills and saws)

7. Toys, leisure and sports equipment (e.g. video games)
8. Medical devices with the exception of all implanted and infected products (e.g. radiotherapy equipment)
9. Monitoring and control instruments (e.g. smoke detectors)
10. Automatic dispersers (e.g. for hot drinks or monies).

We were first faced with huge computers but now they have become more compact and low cost and consequently easily replaced along with our mobile phones – in 2011 we threw away 41.5 million tonnes of electrical goods and that is expected to rise to 93.5 million tonnes by 2016.[56] Now the challenge is how to ensure that more and more components become recycling friendly.

Professor Ming Wong, director of the Croucher Institute for Environmental Sciences at the Hong Kong Baptist University told the CleanUp 2013 conference in Melbourne that e-waste was 'a global time bomb'. He said it was 'the world's fastest growing waste stream, rising 3 to 5 per cent every year and only a small fraction of this is safely disposed'.[57]

More companies are talking about the issue and how to build in an end of life component to their products but the problem is the greed factor as much as the recycling factor. This has resulted in the smuggling of e-waste to countries simply for its revenue potential rather than the environmental element because these products contain precious metals. India for example banned e-waste imports in 2010 but thousands of illegal shipments still arrive every year. People in the so-called 'informal sector' have tried to break it, burn it or use acid baths to separate metals. Rather than disposing of the waste safely, people are resorting to any means possible in order to access the gold, silver and platinum, at great risk to their health from the likes of arsenic, lead, mercury and cadmium contaminants. In addition the vast majority of people in developing regions such as Africa cannot afford new electrical goods so there is a huge market in second-hand products.

E-waste is going to be an area of major growth in the next 10 years as more and more people are becoming computer and telephone literate. Many white goods also have to be handled with care; even the humble washing machines have electronic systems along with the metal and plastic components.

Although more stringent controls have reduced the amount of illegal shipments to countries particularly in Africa, these developing nations are themselves throwing out computers in their hundreds of millions exceeding even the developed nations as they adopt newer technologies.

However, as I have suggested throughout, the vital issue of quality is having an impact on what we can recycle and what we can export. As a result of these

56   Margaret Bates, University of Northampton, 19 August 2013
57   Anna Salleh, ABC 16 September

increasingly stringent import controls combined with the legal requirement for increased recycling domestically, there is a problem which cannot simply be resolved by more of the same. What local authorities have been doing is passing the responsibility on to the collectors without considering how the collectors are going to meet the new quality specifications of the buyers of their product.

There is only one certainty: producers of waste – you and me as well as industry – along with the collectors will have to make sure that the waste stream can produce the profile of recycled waste which has a recognised market. In many countries we still find the tendency to collect and bale material as it comes from the councils and directly export the recyclable which is general of poor quality. This regime is going to stop.

More and more legislation is being passed down the supply chain whereas the legislation should be directed at those who are collecting and are responsible for the waste. Households, supermarkets and industry are the ones generating the waste and local authorities are collecting via their contractors. Local authorities really do have a duty of care and responsibility to make sure that the waste meets a certain specification in order to be identified as a raw material and no longer just as rubbish. That is the challenge and people are washing their hands of the responsibility up the chain when down the chain waste has been collected in a format which is unacceptable because not enough time and money has been put into infrastructure and into collecting and sorting that waste in an orderly fashion so that less time and money has to be spent further along the line where it has to be resorted and classified into grades which qualify as a raw material.

Waste is a raw material, yes, but if it is not going to be collected in the first place in the right fashion – i.e. single stream versus multi-stream – we are failing. Single stream may have economic advantages as we discussed earlier but in the next ten years will single stream be the right answer? Will it help increase the collection of waste and usage of recyclables by the reprocessor and recycling companies? The challenge is how to improve the collection on the doorstep in such a way that as much segregation as possible can be done to help increase recyclability. For industrial collection it is important that the sources where the waste is generated – supermarket, industries or printers – also segregate their waste in a reasonable quality format. For example at the supermarket a bale of cardboard boxes should not also contain leaflets and pamphlets; the cardboard bale should not contain plastic, nor should a cardboard bale contain waxed and wet strength boxes. Boxes used for frozen foods tend to be of wet strength which requires extra processing time. But all these things mean cost and I can see the supermarkets and the ordinary householders throwing their hands up and saying we are already doing enough.

The effort all depends on the waste stream we are discussing. If we are talking about precious metal waste, which we are not collecting in large volumes, we can probably spend more time in segregating them because there is a revenue advantage. Then there are the other mid-stream items such as the non-ferrous metal aluminium cans, copper and brass, those items again have a high recyclable value and low volume of collection compared with other grades such as waste paper. There again we can probably help with more investment in technology for metal segregation to ensure aluminium cans are not contaminated with steel hairspray cans.

But when we come to the low end, high volume of the market – waste paper and plastic – we are restricted by the price the mill can pay. Contamination rates are much higher in waste paper than metals so the challenges at the lower end are becoming tougher.

Quite simply, if we don't have the right product there is no market downstream – but that argument I know is not enough to stop us using paper. The answer will require some brave thinking on the part of planners, local authorities and governments to adopt a radical solution. But before we consider some solutions let us first take a closer look at alternatives to pure recycling and landfill – waste as a source of renewable energy.

6

Waste-to-energy

Waste-to-energy (WTE) is where science meets the bins and the baling, the shaking and the sorting of the recycling industry. It is the 'glamorous' side of the sector which attracts the new investors all anxious to be in at the start of the next big discovery. WTE is the term used to describe the conversion of waste by-products into useful steam, steam-generated electricity or 'synthetic-fuels'. Typically, WTE is produced by treating municipal solid waste (MSW), which is defined as residential and community refuse; the biomass element in MSW includes organic waste (putrescibles), paper and plastic and makes up the largest source of waste in industrialised countries – around 80 per cent. It is also very big business. 'The global market for biological and thermochemical waste-to-energy technologies is expected to reach US$7.4 billion in 2013 and grow to US$ 29.2 billion by 2022,' said Salman Zafar of Ecomena consultancy.

The WTE facilities are an essential, even innovative, part of the waste story but they are not the whole story and like every other facility they need to be fed. They are like delicate plants, they cannot be given any old feedstock, they are quite choosy and won't accept certain textiles or certain timber and, like all exotic creatures, they are expensive. So not all waste can be sent to WTE facilities because it still needs to be treated to be brought to a grade suitable for a WTE company and depending on the investment they have made they will order different types of material for their own purposes.

It takes a leap of investment faith to build a WTE plant – some have had their fingers burned and the plants forced to shut down either because the technology has failed to deliver or because there was not enough feedstock to satisfy the demanding appetites of the furnaces.

Despite the risks, WTE is an important element, a necessity, in how we handle all our waste; my only caveat is that it should not be treated as the first port of call. In fact WTE should be the last resort just before we opt for landfills not the first option when we are trying to reduce our waste mountains. The first goal should always be to convert recyclable waste into recycled products. If we put WTE at the start of the chain we are losing valuable commodities which are useful for creating the very products for which we are going to use the energy. If we don't recycle them and just create the energy then we have missed the recycling process and we are back to the original raw material – trees for pulp, bauxite for aluminium – and we are actually going to increase the quantum of energy we need because we have burned the products instead of recycling them.

WTE is not far flung or fanciful technology. Europe, the largest WTE technology market in the world, treats 95 million tonnes of MSW and commercial waste at some 520 WTE plants each year providing electricity to millions of people.[58] The industry has also been producing heat and power in the United States for a century, and there are some 100 WTE plants in the US alone; China is said to be aiming for 200 facilities by 2020.[59] More recently, however, the definition of waste has been expanded from MSW to include wastes such as wood, wood waste, peat, wood sludge, agricultural waste, straw, tyres, landfill gases, fish oils, paper industry liquors, railway ties, and utility poles.

WTE is essentially a form of energy recovery using physical, thermal and biological technologies. Most WTE processes produce electricity directly through combustion or produce a combustible fuel commodity, such as methane, methanol, ethanol or synthetic fuels.

There are many benefits in using WTE including preventing the release of greenhouse gases such as methane into the atmosphere if the trash were landfilled. It reduces the impact on landfills by cutting the volume of the waste by 80 to 90 per cent, provides an alternative to coal use which prevents the release of emissions such as nitrogen oxides into the atmosphere and saves the Earth's natural resources by reducing the use of oil, coal, or natural gas for electricity generation.

So how does it work? WTE facilities can be divided into some general process types: mass burn, refuse-derived fuel (RDF) and biological.

Mass burn facilities process raw waste that has not been shredded, sized, or separated before combustion, although large items such as appliances and hazardous waste materials and batteries are removed before combustion. In mass burn systems, untreated MSW is simply burned, with the heat produced converted into steam, which can then be passed through a steam turbine to generate electricity or used directly to supply heat to nearby industries or buildings.

RDF is a result of processing MSW to separate the combustible fraction from the non-combustible, such as metals and glass. It is mainly composed of paper, plastic, wood, and kitchen or yard wastes, and has a higher energy content

---

58   Waste-to-energy 2013/2014 – ecoprog
59   *Waste Management World*

than untreated MSW. Like MSW, RDF can be burned to produce steam and/ or electricity. A benefit of using RDF is that it can be shredded into uniformly sized particles or compressed into briquettes, both of which facilitate handling, transportation, and combustion. Another benefit of RDF rather than raw MSW is that fewer non-combustibles such as heavy metals are burned.

Thirdly there are processes using bacterial fermentation to digest organic wastes to yield fuel such as anaerobic digestion.

The inputs and outputs of a typical WTE facility operated by up to 50 people are summarised in Figure 6.

- Typical capacity: 60,000–600,000 tpa (although typically 100,000– 250,000 tpa)
- Land requirements: typically 2.5–3.5 ha

Capital and operational costs for such plants will generally depend on the capacity.

Figure 6 Inputs and outputs of a typical WTE facility.

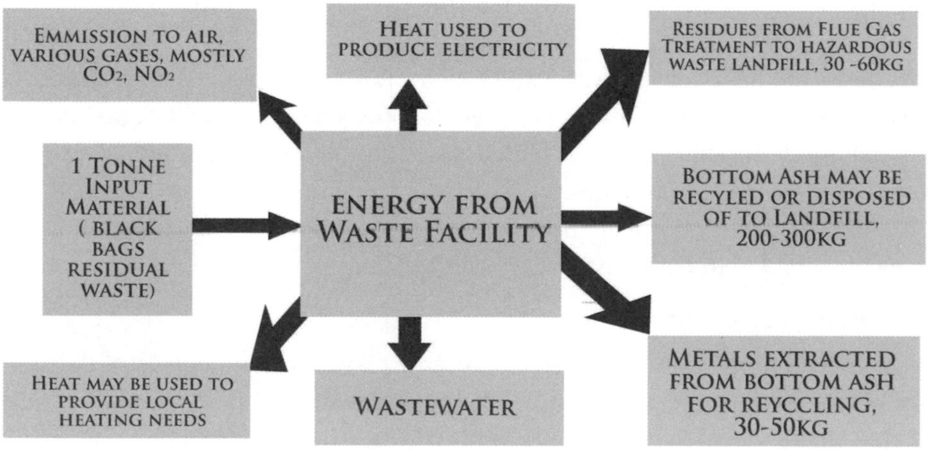

Let us review some of the technologies.

Thermal treatments

The first and probably the most controversial because of its historical reputation is incineration. This is essentially a waste disposal method that involves combustion of waste material. Incinerators convert waste materials into heat, gas, steam and ash. Incineration is used to dispose of solid, liquid and gaseous waste. It is recognised as a practical method of disposing of certain hazardous chemicals such as biological medical waste. When combined with energy recovery it is also the most common waste-to-energy method used. The idea of burning our waste has been around for many years – the first municipal incinerator began operating in the UK in the 1870s and by1912 there were more than 300.[60]

However, incineration is a controversial method of waste disposal due to issues of gaseous pollutants. Particular concern is focused on some very persistent organics such as dioxins which may be created and which may have serious environmental consequences in areas around the incinerators. Despite public fears modern incinerators have high temperature furnaces (*c.*2000°F) which are very effective. The harmful gases which in the past would have quite literally gone up in smoke out of the chimney are now harnessed not only to produce energy to run steam turbines but also to make products such as feedstocks or even diesel for cars by converting the gases produced. It is now a sophisticated process and bears no relation to the traditional furnaces.

Denmark and Sweden have been leaders in using the energy generated from incineration for more than a century. A number of other European Countries rely heavily on incineration for handling municipal waste, in particular Luxembourg, The Netherlands, Germany and France.

Every 100,000 tonnes of waste incinerated can produce 7 MW of power which can keep the lights on in over 5,500 houses for a year. Nevertheless concerns remain about the residual ash left over and the toxins it might contain even though modern facilities have in-built devices, cleaners and filters and the ash left over is within safe limits. However there is nothing that can be said to persuade someone of the benefits of an incinerator if one is about to be built next door but the comparative strengths and weaknesses are summarised in Table 5.

Pyrolysis and gasification

Like incineration, pyrolysis, gasification and plasma technologies are thermal processes that use high temperatures to break down waste. The main difference is that they use less oxygen than traditional mass-burn incineration. However, they are still classified as incineration in the European Union's Waste Incineration Directive and have to meet the mandatory emissions limits that it sets.

These technologies are sometimes known as advanced thermal technologies or alternative conversion technologies. They typically rely on carbon-based waste such as paper, petroleum-based wastes like plastics, and organic materials such as food scraps. The waste is broken down to create gas, solid and liquid residues. The gases can then be combusted in a secondary process.

The pyrolysis process thermally degrades waste in the absence of air (and oxygen). Gasification is a process in which materials are exposed to some

---

60   *Waste Treatment and Disposal* by Paul T Williams (John Wiley & Sons)

Table 5

Strengths	Weaknesses
1. Handle MSW waste with no pre-treatment required	1. High capital costs. Since fixed costs are high the need for consistently high utilisation is paramount.
2. State-of-the-art technology in global use including pollution control technology	2. Negative public perception – NIMBY (stack emissions and lack of understanding of technology)
3. Energy recovery including Combined Heat and Power (CHP) plants and opportunity for district heating programme.	3. Residue quality and disposal, although bottom ash can be reused
4. No long term liability	4. Debate over measurement and long term health effects of dioxin emissions, it should be noted that controls issued throughout the 1990s and more recently with the Waste Incineration Directive have reduced dioxin emissions to well below that of other combustion processes.
5. Proven and commercially available technology	5. Minimum materials recovery, except for ferrous materials
6. Reduces volume of waste by 90 per cent	6. Minimum or guaranteed tonnage may be required by the operator to cover costs

oxygen, but not enough to allow combustion to occur. Temperatures are usually above 750°C. In some systems the pyrolysis phase is followed by a second gasification stage, in order that more of the energy carrying gases are liberated from the waste.

In plasma technologies the waste is heated with a plasma arc (6,000–10,000°C) to create gases and vitrified slag. In some cases the plasma stage may follow on from a gasification stage. The multiple stage plasma operation begins with feedstock which can include hazardous waste; it is dried and sorted, followed by gasification using the plasma torches. The gas is 'cleaned' and heat exchangers recycle the heat created and finally fuel as well as hydrogen and chemicals are produced.

The main product of gasification and pyrolysis is syngas, which is composed mainly of carbon monoxide and hydrogen (85 per cent), with smaller quantities of carbon dioxide, nitrogen, methane and various other hydrocarbon gases.

Syngas has a calorific value, so it can be used as a fuel to generate electricity or steam or as a basic chemical feedstock in the petrochemical and refining industries. The calorific value of this syngas will depend upon the composition of the input waste to the gasifier.

It is worth giving a quick comparison of the advantages and disadvantages of pyrolysis and gasification over incineration. In the first place by using less oxygen than incineration, fewer air emissions may be produced. However, if the gases and oils coming off the process are then burnt, this may also generate emissions; sometimes technology promoters do not make this clear. Pyrolysis plants are modular. They are made up of small units which can be added to or taken away as waste streams or volumes change and are therefore more flexible and can operate at a smaller scale than mass-burn incinerators. This is an important benefit particularly at a time when we are now facing an increase in residual waste.

Pyrolysis and gasification plants are quicker to build than incinerators and it is claimed that their processes produce a more useful product than standard incineration – gases, oils and solid char can be used as a fuel, or purified and used as a feedstock for petro-chemicals and other applications. The syngas may be used to generate energy more efficiently, if a gas engine (and potentially a fuel cell) is used, while incineration can only generate energy less efficiently via steam turbines.

On the other hand gasification and pyrolysis share some of the same *disadvantages* as mass-burn incineration. They only deal with truly residual waste (what is left once maximum recycling and composting has happened). Furthermore plants need a certain quantity of particular types of materials in order to work effectively: for example, plastic, paper, and food waste. However this conflicts with recycling and composting as these materials are often the most valuable parts of the waste stream for these processes.

In addition any fuel produced will not make up for the energy spent in manufacturing new products so reuse and recycling are still better because, like incineration and landfill, energy savings from waste prevention and recycling are likely to be greater than the energy produced. The disposal of ash and other by-products may still be required no matter how superior the ash may be when compared with incineration ash.

Non-thermal technologies

We should also consider some important non-thermal technologies such as anaerobic digestion (AD). This is a biological process that happens naturally when bacteria break down organic matter in environments with little or no oxygen. It is effectively a controlled and enclosed version of the anaerobic breakdown of organic waste in landfill which releases methane.

Almost any organic material can be processed with AD, including waste paper and cardboard (which is of too low a grade to recycle, e.g. because of food contamination), grass clippings, leftover food, industrial effluents, sewage and animal waste.

AD produces a biogas made up of around 60 per cent methane and 40 per cent carbon dioxide (CO_2). This can be burnt to generate heat or electricity or can be used as a vehicle fuel. If it is used to generate electricity the biogas needs to be scrubbed. It can then power the AD process or be added to the national grid and provide heat homes.

In the UK, AD has until recently been limited to small on-farm digesters. However AD is widely used across Europe. Denmark has a number of farm co-operative AD plants which produce electricity and district heating for local villages; biogas plants have been built in Sweden to produce vehicle fuel for fleets of town buses; and Germany and Austria have several thousand on-farm digesters treating mixtures of manure, energy crops and restaurant waste, with the biogas used to produce electricity. AD is also widespread in other parts of the world. India and Thailand have several thousand mostly small-scale plants. In developing countries, simple home and farm-based AD systems offer the potential for cheap, low-cost energy from biogas.

Anaerobic digestion provides an important opportunity to generate 100 per cent renewable energy from biodegradable waste. Research clearly indicates the most sustainable way to treat our food waste is to have separate weekly collections for treatment by AD.

Landfill, fermentation and mechanical biological treatment

Disposing of waste in a landfill of course involves burying the waste, and this remains a common practice in most countries. Landfills were often established in abandoned or unused quarries, mining voids or borrow pits. A properly designed and well-managed landfill can be a hygienic and relatively inexpensive method of disposing of waste materials. Older, poorly designed or poorly managed landfills can create a number of adverse environmental impacts such as wind-blown litter, attraction of vermin, and generation of liquid leachate. Another common by-product of landfills is gas (mostly composed of methane and carbon dioxide) which is produced as organic waste breaks down anaerobically. This gas can create odour problems, kill surface vegetation, and is a greenhouse gas.

Ethanol, lactic acid and hydrogen are examples of the products of *fermentation*. In the recycling process fermentation can produce ethanol which may be used as a biofuel – a fuel that can be mixed with petrol or used alone and is an alternative to burning fossil fuels. It can heat water to produce steam which in turn drives turbines and is deemed 'carbon neutral'.

Mechanical biological treatment (MBT) is an integration of different mechanical and biological processes which may be incorporated in one plant. The flexibility of an MBT plant is reflected in this DEFRA list of possible uses:

- pre-treatment of waste going to landfill;
- diversion of non-biodegradable and biodegradable MSW going to landfill through the mechanical sorting of MSW into materials for recycling and/or energy recovery as refuse derived fuel (RDF);
- diversion of biodegradable MSW going to landfill by:
  - reducing the dry mass of biodegradable municipal waste (BMW) prior to landfill;
  - reducing the biodegradability of BMW prior to landfill;
- stabilisation into a compost-like output (CLO) for use on land;
- conversion into a combustible biogas for energy recovery; and/or
- drying materials to produce a high calorific, organic-rich fraction for use as RDF.

Refuse-derived fuel (RDF) or *solid recovered fuel/specified recovered fuel* (SRF) is a fuel produced by shredding and dehydrating MSW in a converter or steam pressure treating in an autoclave. RDF consists largely of organic components of municipal waste such as plastics and biodegradable waste. RDF processing facilities are normally located near a source of MSW and, while an optional combustion facility is normally close to the processing facility, it may also be located at a remote location.

To summarise then, WTE technologies hold the potential to create renewable energy from waste matter, including municipal solid waste, industrial waste, agricultural waste, and industrial by-products. Besides the recovery of substantial energy, these technologies can lead to a significant reduction in the overall waste quantities requiring final disposal, which can be better managed for safe disposal in a controlled manner. They contribute substantially to greenhouse gas mitigation through both reductions of fossil carbon emissions and long-term storage of carbon in biomass wastes. The most modern WTE systems options offer significant, cost-effective and perpetual opportunities for greenhouse gas emission reductions. And of course they create employment in rural areas, a reduction of a country's dependency on imported energy carriers (and the related improvement of the balance of trade), better waste control, and potentially benign effects with regard to biodiversity, desertification and recreational value.

WTE is not only a solution to reduce the volume of waste and provide a supplemental energy source, but also yields a number of social benefits that cannot easily be quantified. Every year there are further innovations from turning stray forms of energy such as noise or random vibrations from the environment into useful forms of energy, developing electricity from materials put under mechanical stress – the piezoelectric effect – and extracting biodiesel fuel and low grade animal feedstock from chicken feathers; a useful by-product from the 11 billion lbs of poultry industry waste that accumulates annually.

Although the growth of WTE facilities in Europe has been strong in the past decade, that growth is expected to slow as investment tails off as a result of a combination of the post-2008 economic slump with continuing difficulties in Southern Europe and the building boom of the 2000s. There is an over capacity of WTE facilities in Europe – at least for the next five years. This hiatus in construction should pass as Eastern European countries begin to implement European Directives and the economies of countries such as Spain and Italy begin to recover resulting in more waste, particularly from commercial sources. Thereafter it is anticipated that there will be more but smaller and more versatile operations.

The recycling industry is a constantly changing business supported by innovative and dynamic developments. Some of the new projects in France alone include looking at turning fatty wastes into solid pellets for agricultural fertiliser, turning tennis balls into exercise mats and even extracting benefits from Nespresso capsules. We need new solutions and there will always be innovative technologies being developed. Hopefully in the future some WTE systems will be incorporated for domestic usage – but they should not be regarded as the only solution. The answer will be to develop an integrated approach to disposing of MSW making the best use of every methodology at our disposal. In time we will all have to accept this new responsibility; the only question is: when will consumers recognise their obligations?

7

The role of consumers

We are all consumers; therefore we are all producers of waste which means we all share that burden of responsibility to work towards a satisfactory solution of the problem. Let's just remind ourselves of the scale of the challenge. In 2011 Americans generated about 250 million tonnes of waste according to the Environmental Protection Agency and recycled or composted about 87 million tonnes of it – that's 1.5 lbs from every 4.40 lbs per day per person. Those figures exclude industrial, hazardous and construction waste and are just what is thrown out of homes: everything from food and packaging to computers and sofas. By comparison in Europe in 2010, the total generation of waste from economic activities and households in the 27 member countries amounted to 2,502 million tonnes; so inhabitants generated on average about 5.0 tonnes of waste each. In the UK for example MSW peaked at just over 36 million tonnes in 2004 and declined steadily to 32.4 million tonnes in 2010.[61] In Asian countries, the urban areas produce *c.*760,000 tonnes of MSW per day and this is expected to rise to 1.8 million tonnes per day by 2025.[62]

Surpassing all others is China. According to a World Bank report China is likely to produce twice as much waste as America by 2030.[63] But we are all in this together and the report calculates that collectively the inhabitants of the world's cities – estimated at around three billion people – generate 1.3 billion tonnes of solid waste per annum which the authors say is likely to reach 2.2 billion by 2025.

No one doubts the scale of the problem but what are we prepared to do about it? First, instead of assigning blame we must create greater awareness through education. Those who make the products we buy are just as responsible as their customers. The responsibility should not end when the customers walk out of the door with their purchases.

Experiments have been tried with a programme called Extended Producer

---

61   Eurostat 2012
62   World Bank 1999
63   *What a Waste: A Global Review of Waste Management* – World Bank

Responsibility (EPR). The idea is that the product manufacturer must share some of the burden when the customer no longer wants products such as batteries, computers and paint. The principle is that it should not be up to local authorities to pay for the safe disposal of sometimes hazardous waste but that the manufacturers should also be a partner in setting up specified collection points or even make their products more robust so they last longer.

Inevitably cost becomes an issue: how much extra are customers willing to pay if manufacturers choose to pass the additional burden of collection on to them? On the other hand some manufacturers see this as an opportunity to develop by displaying their green credentials and establishing pick-up points.

There should be a revenue contribution from the product manufacturer but this is very difficult to govern. Take cardboard boxes for example: as we have noted the UK collects just under ten million tonnes of waste paper so the question then is how much of that waste paper collected is made from boxes manufactured in the UK, how much has come from European companies and how much from companies outside the EU? Therefore how can European or Asian box maker be penalised? Or should the importers who bring the box into the country be penalised? In Europe as there is free trade of goods it would be impossible to know whose boxes are going where. So while there has to be some degree of responsibility on the part of the manufacturer at the moment it is difficult to see how it could be fairly implemented.

Awareness of this type of initiative involves a process of education and this should be a common theme of all recycling. Everyone should be aware of the consequences of what they are doing when they choose to dispose of anything. We are already used to multiple bins and multiple collections, so can we be persuaded to take our redundant computers to, say, computer-specific collection points? Shawn Lavin of Pioneer Industries International, Minneapolis, Minn., said the hauling and recycling industry needed to fill a large void in recycling knowledge to create more effective recycling programmes. He for one was surprised there was not more policing of collections.[64]

If the message cannot get through on a voluntary basis then perhaps the authorities will have to rely on compulsion. It will become illegal simply to throw our used batteries into the general waste stream; we may be charged as they have been in Switzerland for specific bags of waste and the contents of those bags can be scrutinised. Will we go to the extreme of removing every identifying label from our bags of rubbish in case a 'waste inspector' can track us down from the contents of our own bag? Such 'Big Brother' snooping will cause outrage; so too will additional charges on our waste collections – people will be reluctant to pay higher local taxes.

---

64   *Recycling Today,* 16 October 2013

But let's look at it another way – why should we be allowed to dump our waste without a care? The whole recycling success story is entirely dependent on the waste generator who has a duty thereafter to ensure it is used most economically for the benefit of the planet. I would not like rubbish to be thrown into my bedroom; in the same way we are sleeping on this planet, generating waste on this planet, we are duty bound therefore to present it in such a way that we reduce the toxicity and other side effects of the waste that is accumulating in mountains around the world.

Perhaps a more compelling argument would be to ask ourselves how much waste do we think is acceptable: none at all or just a little; a small amount of litter blowing down our streets, just a few un-emptied dustbins, just a modest amount of effluent in our rivers and seas. 'Waste not, want not', we are chastised as children but we soon forget the admonition when we grow up to be adults even though we know that by wasting anything we are being inefficient.

We should first say that all the people in the waste recycling supply chain are the real green ambassadors. What they are doing is contributing to extending the life of, for example, the fibre of paper. Every piece of paper, or for that matter other waste product, can be recycled once or twice, fibres have been recycled five or six times. But that only works if that fibre, from the time it is generated as waste, is looked after and cared for so that the manufacturer can use it as a raw material otherwise it is wasted and thrown in landfill or burned.

One of the important roles the consumer plays is safeguarding this from happening. Every consumer to me is a true green ambassador. Maybe not enough is being done to thank them so I would like to recognise everyone who is reading this book and express my gratitude. However, different systems of collection are putting different pressures on the consumer whether it is the Swiss putting tags on the rubbish or the mills requiring the complete footprint of every bale that goes to the mill for processing, insisting on a whole history: where did the paper come from, how was it collected and how was it sorted?

More important is the lack of understanding by householders about why and how they are supposed to segregate their waste. So while we congratulate the householders we also need to educate them about why they are doing what they are doing. For example if all the waste paper goes in one pile we will end up collecting the lowest quality of waste paper which might contain newspaper, office stationery and some plastic contamination. If all these things are put in one bag it becomes a single stream; the different ingredients in the pile of paper have completely different usages to produce different end products. Newsprint and magazines are primarily reused for newsprint; some of it is used in the container board industry to give density to the paper board while the white paper can be used to create another white paper, which is expensive, or tissue paper.

Of course one has to be practical about the extent to which the consumers at household level can segregate their waste. For example ideally with plastic milk bottles should be separated from PET bottles, or separated from shrink wrappers or plastic films. Not only do they vary in price but they also vary in uses. Naturally the number of milk bottles that a household uses in a week does not warrant that level of separation but it illustrates the difference even within what we loosely call plastic waste.

How much segregation a household should be expected to do is the challenge: the very basic separation is to divide wet and dry waste. Food and garden waste should not be mixed with dry recyclables. Those dry recyclables might consist of plastic, aluminium cans and glass.

Glass contaminated paper is only going to end up damaging the paper machine so a paper manufacturer is not going to buy waste which might be contaminated with shards of glass. The next stage would be to divide glass into different colours, which of course a householder cannot be expected to do. But the separation done at household level needs to optimise the usages of these materials and as a result produce a better grade of recyclable that helps to save energy, reprocessing costs and reduce carbon emissions during the production cycle.

To a great extent modern technology helps. Within paper recycling, while it makes sense to separate newspaper and other paper, with modern facilities we are able to separate cardboard from newspaper so it is acceptable for them to be mixed together. Typically all plastic – milk bottles, soft drinks bottles – is mixed together but again the technology helps. At the recycling plant plastic milk bottles of a natural colour are separated from the PET bottles which tend to be mixed colour. There is even technology available to segregate different colours using lasers and other devices. The ultimate of plastic recycling is where the consumer can gain value by segregating all these colours and grades, but the consumer needs to be educated about HDPE and PET. At the moment we are collecting plastic as one grade and leaving the plants to do the sorting.

In other words it is not all 'just paper' or 'just plastic'. The aim must be to help the consumer to understand why it is important to do what they are doing. There is a need for local councils or local collectors to give more information to the householders. At best for now households should endeavour to separate the basic broad categories of waste.

Western economies are all strongly promoting the green concept: Switzerland, Germany, Scandinavian countries, the UK, they are all constantly trying to engineer a cost-effective way of sorting waste at both the household level and at a commercial level. However there is no standardisation as to how it is being sorted. Volume of waste dictates the method of collection. If an area only has a population of 20,000 people spread over a 10-mile radius then the density of waste per mile is less and therefore the generation of waste will be less. Whereas if in an area with a 10-mile radius with 70,000 people the waste will be greater and therefore the collection strategy in both locations will naturally be different – one size does not fit all which is why blanket edicts for example from Europe can never satisfy every demand and may even hinder the whole recycling effort of a neighbourhood.

By contrast, in different emerging economies there appears to be no proper sorting of waste which often ends up in landfill. That is where efforts should be targeted in the next 15–20 years: saving that waste from going into landfills or on to the mountains of unsorted waste. That is the immediate challenge.

The problem in recycling is really that the people who are implementing the recycling decisions are all so remote from the users of the recycled material. They also need to learn and understand what is viable and how far recycling on the doorstep can be promoted. As far as the householders are concerned we should endeavour to help them understand that although they are not receiving any financial incentive every kilo of waste that they help sort amounts to hundreds of millions of kilos across the world which is enormously beneficial to the planet. If they do not do this, in the next 25–30 years our carbon emissions will skyrocket to such high levels that the price of raw materials will go up and in turn consumers will end up having to pay a higher price for their goods. Without recycling, aluminium cans which we are so used to and happily scrunch up and cast aside would all have to come from bauxite which we would use so fast that demand would outstrip supply, forests would have to be cut down for pulp and again prices would escalate and we would end up paying higher prices for the books that our children read at school.

In other words while there is no direct remuneration there is a lot of indirect benefit from the kilo of waste that the householder has helped save and that is eventually going to help everyone else. So education and understanding are critical. We must not be over demanding on the consumer. In many countries we may even be going too far when it comes to the recycling regime. The first step should be to explain what is required and what is in effect the breakeven point between technology and the consumer: how much should the household be able to do without feeling the strain. The technology should come in the sorting system to allow us to extract the maximum benefit from the household collection. Stories circulate that collectors, for example, refuse to empty bins because in the glass container there were two pieces of paper. That kind of extreme control is very frustrating to the householder and does nothing to help the drive for greater recycling and understanding. It is fair to ask the householder to do a certain degree of sorting which has to be better than the sorting we were once used to but not to go too far while allowing the technology to grow so we reach a happy medium. We are making good progress; we have some advanced sorting systems now which cost a great deal but they are also able to do a great deal of segregation.

As always it comes back to economics: the investment in a modern system only makes sense if over the anticipated lifetime of the facility it receives a specific volume of waste – its feedstock. As I see it technology will always go on getting more sophisticated and that can only mean it will require ever larger volumes of waste. In order to achieve that there will have to be a network of larger collecting units across the country feeding these bigger sorting and recycling facilities – an idea I will expand on later.

There are some places like Germany where they are returning to the very old idea of paying a deposit on a plastic bottle so if the consumer returns the bottle he or she gets the deposit back. America also has deposit schemes in some cities such as the Staples Rewards programme in Massachusetts for customers returning toner and ink cartridges; Staples say they have recycled 182 million lbs of plastic and metals since launching the programme in 2005.

There is an effort to promote the idea in the UK among small shopkeepers who can collect the PET and HDPE plastic containers very easily and the volume of bottles is much higher than at householder level making it practical to segregate. This is the first, lowest point of the supply chain where it can be done in higher volumes than in households where not just plastic but also other high-value commodities are not sorted correctly. Supermarkets could also do it very easily and in the most cost-effective way. To be successful we have to reduce the amount of labour and time spent transporting, segregating and recycling. The price of recyclables is governed by the price that consumers are prepared to pay for the end product which in turn limits the price for used plastic bottles. As always the role of the consumer is key but even consumers need to change their habits.

We collect food waste which can be treated by bio-digesters and anaerobic processes. It can be converted into gas or compost. For this to work food waste needs to be collected separately but there may not be sufficient volume to warrant handling it in a cost-effective manner. The real answer to our food mountain is for us to consume less. In Western economies we tend to buy too much of everything without being able to consume it. It would be wise to think twice about the quantity of food we are buying. The food waste message must be: reduce food waste because it is difficult to process and much ends up in landfill. If we reduce food waste it will help the economy by exporting the excess to where there is a demand for it thereby generating revenue and cutting out the landfill cost.

There is no real cultural difference between consumers in different countries when it comes to recycling. A hundred years ago recycling was a way of life for some people. It was a survival job. More recently we have realised that we need this material out of necessity not luxury, so in the 1970s and 1980s recycling took off. It then became apparent that there was a huge business opportunity as the volume of collected waste grew, industry developed, jobs were created and large revenues were generated. Companies boasted of their green credentials and the global trade in waste expanded; today it has a market value of hundreds of billions of dollars.

So if we are to save the planet's finite resources we all have a contribution to make; in my opinion it is not by compulsion but by education that this will be achieved. The consumer cannot be coerced or compelled into helping but by understanding some simple principles of the recycling process that those in the industry understand so well, the householder can make his or her own valuable contribution. As for the industry, it should not take for granted the provider of the raw material which feeds their facilities. And local authorities who manage the collections or even the governments who devise new laws and impose new directives should not assume that those at the bottom of the recycling chain (the householders) necessarily agree with everything being imposed upon them. Recycling will work through education not imposition.

It would be wrong to talk about the habits of consumers without mentioning supermarkets, the source of possibly our greatest amount of waste. The choice on the shelves is bewildering – 38,000 different lines is not uncommon – all in colourful packages and tempting displays.

British families are typical of people all over the world and according to a survey by the Waste and Resources Action Programme (WRAP) Britons throw away £60 in food waste or roughly six meals a week. Bread, milk and potatoes top the list of products regularly discarded along with some 86 million chickens every year.

WRAP's chief executive, Dr Liz Goodwin, said: 'Consumers are seriously worried about the cost of food and how it has increased over recent years. Yet, as WRAP's research shows, we are still wasting millions of tonnes and billions of pounds.'[65]

The UK Government's resource management minister, Dan Rogerson, said in response: 'Cutting waste and driving business innovation will help to build a stronger economy. We will continue to work closely with food retailers and manufacturers to achieve this goal.'[66]

So what is the general experience of a shopper in a supermarket which is doing everything it can to attract our attention to today's 'Best Buy'? Before

---

65   BBC 7 November 2013
66   www.letsrecycle.com

we even talk about food waste we need to consider the packaging. Are we over packaging in our desire to tempt the shopper? And are we offering 'family sized' packs when smaller would do?

The amount of packaging is directly proportional to the amount of waste we create and in my opinion we have to think about reducing it; are we really convinced that an Easter egg is bigger than the next one just because it is wrapped in a huge card box? We are not helping the planet and probably irritating the shopper because when all the wrapping is stripped away only a modest piece of chocolate remains. Does every toy or object have to be so tightly sealed in an impenetrable vacuum pack many times the size of the item itself?

We are said to be confused and fail to understand the difference between 'best before' and 'use-by' dates; perhaps we should simply have smaller plates which would encourage us to have smaller portions.

Perhaps we also need to adjust our expectations and improve our technology. That would be the view of the UK's Institution of Mechanical Engineers (IMechE) which blamed the two billion tonnes of food waste each year on what it considered to be 'unnecessarily strict sell-by dates, buy-one-get-one free and Western consumer demand for cosmetically perfect food'. It also pointed to 'poor engineering and agricultural practices', inadequate infrastructure and poor storage facilities.

There is little doubt that we are driven by the sell-by date. Once an item is past that date it goes into the waste stream further increasing its carbon footprint. Remember those vegetables have already travelled hundreds of miles to reach the shelves and once they go into waste they start a new carbon mile journey. But we all make our own judgement about sell-by dates; those brought up during the Second World War are often scornful of the terrible waste they believe such caution encourages. The manufacturer of the food has a view when making or growing something that by the time the product reaches the shelves it has already been travelling for so many days and possibly many miles. The manufacturer then decides that a product can reasonably be consumed within say 90 days and 90 days minus so many days for travelling gives the sell-by date. But whether it becomes toxic is something each individual can decide. It would seem to make sense not to buy large, 'family-sized' packs of perishable goods but non-perishable items may become cost effective. *Caveat emptor* as they say.

Supermarket chains are perfectly well aware that they are one of the mainstream producers of waste because all the large groups with stores in any country in the world have multiple branches and distribution centres. The amount of waste that any supermarket group is generating put together would come to some hundreds of thousands of tonnes, forming a large quantum of the waste we collect. The waste they produce is predominantly paper, plastic and food waste as well as aluminium and glass. They know this and they know that waste has a value which can spark what I see as healthy competition in waste recycling.

Why shouldn't supermarkets benefit from this value? Any recyclable that is collected, sorted and sold has a value, a value which the paper mill or plastic plant is prepared to pay for this material. So it is in the interest of the supermarket to try to increase the amount it gets from the waste that is generated across the group, at the same time it can pass the benefits on to the consumer. Shoppers can always vote with their feet if they feels they are not getting good value but from an environmental point of view the supermarket is supporting and promoting the recycling companies and helping them improve technology which will ultimately prolong the life of recyclable waste. This has to be a good outcome. Supermarkets in effect can enter a closed loop where the used boxes go to a recycling plant to make more boxes which are sold back.

There is room for improvement, however, and it would be useful to get supermarkets to realise how different grades of paper and plastic down the chain connect with the products into which they will be eventually be converted – the policy of education I referred to earlier. If there were a greater understanding of this chain, supermarkets could help cut the cost of sorting and excessive logistics by collecting in a format that can go directly to, say, a paper mill without having to go first to a sorting station. They might also get more for their recyclable waste.

From a recycling point of view, supermarkets have an important role. They have an excessive amount of unsold newspapers, an excessive amount of cardboard boxes and excessive amounts of office and printing paper but they all have different uses. Properly sorted and baled material free from contamination would immediately help increase the yield that a mill could achieve. A mill that is making newsprint would love to receive bales of newspapers and magazines uncontaminated with cardboard, printing and writing paper made from wood-free pulp. So I am in favour of more and more supermarkets working in partnership with recycling companies to promote a greater degree of recycling but there is a need for supermarkets to learn and to try to spend time segregating. Indeed I am in favour of a global free and fair trade of waste products; if different bodies are fighting over every piece of waste then many of our landfill problems will be solved. Whoever pays the price takes the material and this merely encourages everyone – supermarkets, shopping malls, industry, local authorities – to use their waste as they think best.

Every effort that supermarket chains around the world put into collecting waste means they are at that point becoming members of the global recycling loop. That loop becomes stronger and instead of talking about a closed loop it becomes an open global loop. However let us not be naive. Recycling is a revenue-driven business and when German stores offer to repay a deposit on any plastic bottle returned to them it is for a purpose – a purpose which also happens to preserve the environment. For now people are putting in time, effort and money in order to achieve a higher financial return. The day that balance changes is the day the recycling rate will start to go down.

In their defence supermarkets would point to the efforts they are making in reducing waste. Carrier bags use 70 per cent less plastic than they did 20 years ago; nevertheless this thin plastic also used to wrap multipacks and toilet rolls represents 645,000 tonnes or 43 per cent of all plastic household waste in the UK each year, whereas plastic bottles account for only 32 per cent in comparison.[67]

It remains the case however that supermarkets operating in emerging countries have some way to develop in terms of recycling compared with their counterparts in Western economies. Very few in Africa, Asia and South America are able to collect paper and plastics let alone segregate them and much still goes to landfill, which brings me back to where we should be putting our effort. So much more can be achieved by encouraging the principle of recycling in emerging countries ,where at best recycling levels are no more than 25 per cent, than by trying to pursue the elusive goal of zero waste in the West. There is a long way to go.

We will go on shopping and we will go on being tempted by 'Best Buys' and 'Special Discounts'. We demand fruit and vegetables and exotic products from around the world regardless of season and to satisfy that insatiable demand the supermarkets will go on supplying the goods. We cannot complain about waste mountains, recycling facilities and even incinerators if we always insist on more and throw away what we don't want.

It is not the job of the recycling industry to change human nature; it is our job to manage the waste that is created in the most effective and, yes, profitable manner. It is also our job to alert government about what is happening in the industry and how people's habits are impacting the industry. Supermarkets are a perfect barometer of those changing habits; they reflect how Asian shoppers are eating more meat, how populations are growing and becoming more acquisitive, how emerging nations require more electronics, more white goods, more of everything. The question for every government is, how will this all be managed?

---

67   *Retail Gazette*, 7 April 2011

8

The role of government

There is a difference between government having a role and having a responsibility. Governments are not to blame for waste mountains: people are. Equally governments are supposed to govern which is not the same as interfering with private business but it does mean that they have a duty of care, care for the people and care for their lands, both essential parts of a government's brief. So if governments around the world see their peoples behaving as all humans do, consuming more and more every day and wasting more and more, they have no choice but to act in the absence of any other brake on our relentless pursuit of what we believe we need.

Recycling is the only way forward, we only have to look at towns and cities where the collection service breaks down for just a few days to realise that; and who best to drive the recycling message home than governments. They introduce regulations which need to be enforced so the producers of waste at household and commercial level are made answerable for the waste they create, bringing them into the recycling circle. In short if we refuse to help ourselves someone in authority will have to do it for us.

But legislation can only go so far. For legislation to work the citizens generating the waste should feel like 'green ambassadors'; it is not the legislation itself which should be compelling us to act thoughtfully but the need to do good for the planet by participating in recycling programmes. By working together we will certainly be able to reduce the volume of waste going into landfills and certainly be able to present the waste in a format that can be readily used.

Likewise for the industry itself, the state has long given up hoping for self-regulation and is increasingly imposing ever more stringent laws and directives on how the recycling industry should behave, what can be shipped round the world, how it should be shipped, how it should be inspected and even what actually constitutes waste. The industry doesn't like it but then no one likes politicians 'interfering'. In the UK in 2013 the media objected to a Royal Charter on Press Regulation arguing that it amounted to a curb on that most precious right – press freedom – and it would be better for the press to regulate itself.

The question therefore for the recycling industry is: can it help itself? Yes, it can, but it unquestionably needs to work in closer collaboration with manufacturers of the very products everyone buys and ultimately discards.

First let us look at the quantities of material we are creating. Governments will celebrate successful businesses and happily tax their profits but manufacturers have a responsibility for what they create. They are not allowed to dump contaminated waste into rivers so why should they be allowed to 'dump' huge quantities of material in the form of excess packaging on to society and then walk away. Extended Producer Responsibility, mentioned in the previous chapter and already being tried to a greater or lesser extent in many countries, must be extended to include all waste; the first EPR laws in America were introduced back in 1991 by Minnesota and New Jersey so we should have resolved the teething problems by now; instead we are still merely tinkering with the idea.

Manufacturers, the producers, are not in the recycling business, but they should share in the cost of disposing of waste. If that means that the items they produce – the cars, the food, the toys – become more expensive then that is the price the consumers have to pay, not the general taxpayer. Equally, insisting that producers bear some responsibility for the collection and disposal of packaging will incentivise them to be more economical in the way they present and sell their products with an obvious positive impact on their bottom line.

Governments must be aware of what is happening internationally. There is no question that along with logistics, demand and fluctuating global economies, quality is the dominant issue and with it the tighter controls being imposed.

It is no use bleating that so much of this smacks of protectionism. It is inevitable that increasing volumes of domestic collection will come on stream in countries into which we have traditionally sent so much of our waste and that competition will increase as a result. We have to meet the challenge and in order to do that the recycling industry has to produce the goods the end user demands. That in turn means a wholesale improvement in infrastructure if MRFs are to stay in business. But without some form of financial support many operations will go under unless they can be helped to rebuild and regroup. This is not just a question of subsidising industry, as I will discuss a little later, it is also about helping countries meet recycling targets imposed by Europe and other international authorities. The UK for example was in danger of missing its recycling targets for 2015 and may face a fine of possibly hundreds of millions of pounds which of course would fall on the taxpayer. However, the greater challenge is not so much whether or not we will reach our targets but in doing so will we be producing more recyclable raw material and thereby promoting recycling? We can recycle and produce a recyclable grade but it must be of the right quality and specification otherwise it will have no market and be of little value apart from being used as feedstock for waste-to-energy.

And what if contrary to expectations residual waste is rising while recycling rates are actually levelling off not increasing, as the Association of Directors of Environment, Economy, Planning & Transport (ADEPT), a coalition of UK local authority waste departments, warned. Steve Kent, president of ADEPT, said 'alarm bells' should be ringing in Whitehall, warning that the UK risks being left 'ill-prepared to manage its waste properly'.[68]

Europe leads the way in waste recycling as well as waste production but the European Commission has a difficult task in trying to harmonise its rules and the targets for recycling it would like to impose on member states. It is essential that there should be both harmonisation of laws worldwide and a greater flexibility to accommodate different economic strengths and technical ability. Roy Hathaway, Europe policy adviser to the waste industry trade body, the Environmental Services Association, said:

> In an EU of 28 Member States, there are huge variations in recycling levels, financial resources and political will to change. Put simply, we do not believe that the poorest performers are on track to meet their targets. Ensuring that they do so must be the EU's first priority. But, given the huge variation across Europe, we don't see a way, at the moment, to set new EU targets that are high enough to challenge Member States with good recycling rates but would still be credible in the poorer performing countries.[69]

So what should governments be doing right now? Should they be collaborating in some way recognising that waste is a global problem because it threatens our natural resources and our very planet, or is it going to be a case of everyone for themselves? Emerging nations expect to be able to grow and enjoy the prosperity that Western economies have been enjoying but I wonder if they are helping themselves by imposing barriers and introducing restrictive practices to bolster native industries or are those obstacles counter-productive?

A major step forward would be the introduction of a global standard upon which waste would be traded and applied to different products; Europe, America and Asia all have different and sometimes conflicting specifications. If this business is to grow globally then trading regions should try to come to some internationally accepted definition of product whether it is paper, metal, plastic or steel. The benefits of this would be that everyone would be using a

---

68   BusinessGreen 8 August 2013
69   Letsrecycle.com 9 September 2013

common standard, whether it is the recycling plant, the global trader or the local authorities collecting from households and communities. From this standard the industry will know that they can produce a product which will be globally accepted. I recognise that it will be a difficult task to harmonise these specifications but we should forget individual specifications for every individual country; the entire EU should certainly work to one specification, then the Americans and Europeans could develop a joint specification which would be a huge achievement.

Once a product is defined in this way buyers know what they are buying and can compare quality and prices. At the same time legislators can then reduce the unnecessary requirement for extra evidence which has to accompany every container because once we have defined the product thereafter the controls should end. It should satisfy customs and international trade requirements to move products from country to country and continent to continent without delay.

The reality of course is that nations will always protect their own. But there is an urgent need for a new global initiative to review all the rules which govern the handling, trading and shipment of waste material in all its forms in a bid to come up with a unified plan of action, along the lines of the Basel Convention on the Control of Transboundary Movements of Hazardous Wastes and their Disposal which came into force in 1992. And yet governments may think industry should make its own case and not rely on them. Dan Rogerson, shortly after being appointed the UK's resource minister in 2013, wrote to stakeholders saying: 'We will be stepping back (in April 2014) in areas where businesses are better placed to act and there is no clear market failure. The responsibility for taking work forward will largely rest with the industries concerned.'

This worried many who felt it was very much the responsibility of governments to speak up on behalf of their own industries; waste management companies cannot speak in the European Parliament and believe it is up to ministers to raise the concerns which the industry is facing and to help forge a coordinated waste strategy amidst all the other noise and lobbying which is fighting for their attention.

We can surely begin by agreeing that the human race will go on consuming and we can recognise that there is probably an average level of recycling that a modern society will attain albeit at an ever-increasing volume. Above all we can agree that waste is itself a precious raw material which can be used effectively to supplement finite resources, that it can provide energy in an environmentally sound way and that, when distributed widely and safely, can benefit all. That is a good basis from which to start talking about how we can tackle the problem together.

Unfortunately for the people for whom the regulations are made – the householders or the industries – either governments are finding that they are not achieving what they really want or they are trying to move much faster than the householder is prepare to go. Having a different bag for every possible category of waste, which may be operated successfully in some countries, is quite a far-fetched ambition for the majority. The very first and most basic step in recycling is understanding that waste is recyclable. The second step is understanding the different kinds of waste, the third is trying to separate dry and wet waste recyclables, the fourth is sorting into different categories and the fifth would be to streamline the categories until the stage is reached where there is zero waste and everything from the household is classified and segregated properly. We are nowhere near that level of sophistication in most countries. Either the rush to achieve that is too fast or the governments are finding that their regulations are driving change much faster than is physically possible on the ground making governments enforce more and more legislation which is ultimately counter-productive.

They are all forgetting that waste is not a machine-generated product at the end of a processing line. Every single unit produced is not an exact replica of the one that went before it nor the one that will come after it. Here we are generating a product that is manually and then mechanically sorted from material that is collected from household or industrial sources so each outcome will not be identical. Some of the legislation is trying to drive in that direction with a requirement guaranteeing that the recycled waste that is generated and then sold around the world carries with it an unlimited amount of evidence to ensure that the waste is accepted at the other end by the buyer.

It is perfectly correct that we should have legislation to prevent the shipment from one country to another of containers that are full literally of rubbish. We need to realise that what is useless rubbish for one country will also be useless rubbish for a recipient country. What should be traded globally must be definable and usable raw material.

Under current European-wide legislation it is necessary that the container of waste recyclates produced at a European plant and exported must carry with it detailed documentation for international movement. On top of this there is a requirement to prove that the waste will be used for manufacturing purposes by the recipient. This is an unnecessary additional burden of proof imposed on EU exporters which needs to be satisfied by completing additional forms known in the trade as Annex 7 export procedures. This extra information required from international buyers is a unique piece of European legislation which does not apply for imports from non-EU countries.

Unfortunately this evidence-producing exercise is making it difficult to move waste around the world not only for the exporter but also for the buyer who may question whether it is worth going through the bureaucratic business of proving and re-proving rather than obtaining waste feedstock from other less demanding countries around the world. The buyer's simple requirement is waste of approved specification which we Europeans are endeavouring to deliver. That should be the sole driver of export movement. The exporter must produce a product which the buyer wants. Instead we are governed by strict European legislation requiring extended form filling and evidence gathering rather than promoting exports and a healthy global trade.

Sometimes the weight of legislation is defeating the basic goal which is that every bit of waste that is generated has got a home, is used by somebody to produce a product thereby saving greenhouse gases, finite raw material, energy costs and ultimately the planet.

I find that the legislators are making it so difficult for people to export that it is becoming a challenge to satisfy legislation rather than running a successful trading business. Controlling regulation is important but we must recognise a point at which it ceases to be an asset to promote trade and becomes a liability which reduces trade.

Incentive is often overlooked when those in authority are deciding how best to achieve their aims. It is an incentive to return plastic and glass bottles for a deposit of a few cents. It is an incentive if customers can be told or even see on a chart how much a supermarket chain has recycled in a month and to know how that is keeping prices down; and it is an incentive for households to participate as best they can in any domestic collection scheme if by doing so local taxes can be kept low. It is an incentive to offer emerging countries support to develop adequate facilities to manage their waste which is increasing at a faster rate than in developed nations. A little more carrot and a little less stick from those in authority will work wonders.

We have touched on it before but it is also a government's duty to explain and to educate the people. They want to know why they have to separate their waste into glass, plastic, wet and dry. And if a council suddenly switches to single-stream or multi-stream collections there has to be an explanation. Modern MRFs are highly sophisticated operations, the waste is scanned by lasers, segregated by magnets, trommels and electric eddy flow currents. The facilities don't smell, they can be air-conditioned and the most sophisticated operations encourage visitors and school parties. They are not all wind-blown dumps, seeping toxic chemicals into the ground. Some are of course and they need to be upgraded or replaced; this is an investment which pays dividends in terms of a wider understanding of the whole waste recycling process, not a chore which is only undertaken to meet some directive issued by a remote authority.

The most pressing issue for world governments is the ever-increasing quantity of waste that a growing and ever more affluent society produces. This is not a problem which is going to disappear; it is not a one-off catastrophe which can be managed once and for all; it is in fact an issue which gets bigger with every passing moment. We are all affected in one way or another which is why it needs a unified, international solution and coordinated effort. As different raw materials are produced in different countries there must be a reciprocal effort; we cannot plunder the resources of a nation producing zinc or bauxite any more than we can strip the Amazon Rainforest of its trees;[70] therefore we have to agree on quotas, on levels of extraction, on reciprocal arrangements in exchange for precious material, on rules about treatment and the impact those treatments may or may not have on the environment.

We need to be sure of our facts about incinerators, about waste-to-energy processes and even about the role and efficacy of landfills. If a country decides to ban the use of landfills and refuses to licence any more what does it propose to do with its surplus waste – export it? Why should it become someone else's problem? What is it offering in return? There is always a consequence for our actions and the reality is we must not think of waste as someone else's problem. There is even a consequence for good laws; there was a marked increase in illegal shipments of waste when prices for landfill and tougher shipping requirements were introduced particularly when the recession hit in 2008. Waste crime, including illegal dumping, does not make many headlines but it does exist. There is an environmental impact from exporting just one more container of rubbish waste to Nigeria adding more carbon miles for the untreated rubbish to travel. If a local authority decides to close just one waste 'dump' to save money there is a knock-on effect as people have to drive further to dispose of their unwanted washing machines or garden furniture.

Just as governments have to face up to what could easily develop into a crisis, so too the recycling industry has to acknowledge and start preparing for what may turn out to be a rapid jump in waste streams as the global economy slowly picks itself up after nearly five years of stagnation and recession.

I have mentioned targets but there is a parallel effort which is certainly needed in the UK and probably many other nations, and that is investment. As we strive to collect and recycle more we are failing to match that effort by investing in high-tech infrastructure which will allow the increased waste streams to be converted into readily usable recyclable raw material. We are seeing facilities producing recyclables which do not meet the required quality standard and therefore have no market. As the quality standards and specifications become

---

70   Brazil says the rate of deforestation in the Amazon increased by 28 per cent between August 2012 and
     July 2013, after years of decline. BBC 15 November 2013

more relevant, the cycle will be reversed with the waste going back to landfill or waste-to-energy (WTE). WTE as I have already stressed is not the answer because anything that is burned is lost forever. Burning should be the last, unavoidable, resort. The issue is, from where should the investment come?

Different countries have different programmes but I have yet to find a programme where the governments or local authorities are working in partnership with the recycling plants in helping them to generate recyclable waste which has got a useful lifecycle. We are therefore more and more reliant on the private investor who is trying to come into the market to build large modern plants. But much of this private investment is led by private equity and venture capitalists who have certainly helped in setting up plants but in many cases have not understood the product mix correctly. As a result they have invested in a great plant but it is not producing the great product that the market needs. So we have already seen, and we will see more in the future, plants either having to close down or be modified to meet the requirement of the product line.

I would like to see the money come from the government to help SMEs. Instead of being faced with the potential closure of what might have been a family-owned business running for 50–100 years because they cannot afford to invest in technology that the market now demands, such SMEs would be handed a lifeline by the government. I would favour some sort of direct investment from local authorities in helping those SMEs build better plants and sorting systems so they could produce a product which has a longer life. End of Waste regulations are really asking us to produce a raw material, not waste, so it would be a logical as well as a green investment to back struggling companies which may have been caught out by changing demands or even legislation itself. As I have said, we need more not fewer MRFs.

The question is should the government take on the responsibility of running some of the recycling stations because all these businesses are only viable for a private investor based on investment and return on capital. The plant needs a regular volume of feedstock to process and produce a recyclable grade with a certain life. If these transfer stations or sorting plants don't get enough tonnage and are forced to cover more miles to get the feedstock/waste, it is likely that with increasing collection costs they will not be able to deliver waste of the right quality and the right price. This will undoubtedly mean closing down loss-making plants and having just one or two dumps to support recycling plants in the area.

But is this a job for government? Should they become involved as co-partners? It is a balance between economics and environmental needs. Public–private enterprise could work but it is only viable if the cost of collecting the waste and the cost of running the plant is lower than the selling price of the product. If those sums don't add up then it becomes a non-starter for private enterprise. On the other hand while the investor is only there to make money, the responsibility for the government is to its citizens. In some of those situations where an SME is struggling governments should consider taking on some of the 'green responsibility'. The alternative is that in remote areas where there are no facilities nearby the waste will end up in black bags going into landfills or simply be fly-tipped. In these circumstances, it is likely that the top companies will continue to dominate.

9

The European markets

I started in this industry in the late 1970s at roughly the same time as the whole business of waste collection and recycling was beginning to open up; I was exporting steel scrap and reject steel from European mills and I gradually widened my market into other areas of waste and began shipping recovered fibre to Asia. As the volume increased I started looking at other waste streams where I could expand and forestry products, like steel, were being shipped around the world. Waste paper is what interested me because again like steel we were reutilising our waste which ultimately was for the betterment of our environment, helping to preserve natural resources. In those days America was the big market in exporting waste paper and the only paper being collected at the time was commercial waste – supermarkets, printers, newspaper publishers and factories. The community collection programmes had not really started in a major way.

In the early 1980s I began exporting waste from the UK to Asia and in those days we focused just on cardboard boxes from the UK to Thailand and then computer printed stationery to India. It is hard to imagine a trading world without them but containers were not commonly used back then and I believe I was one of the first if not the first to export a container of waste paper out of the UK. Europe was more introverted and most of the commercial and industrial tonnage was traded within Europe; the UK used to export as far as Germany, Holland and France, while Germany used to sell to the UK. Exporting to Asia was not the thing at the start of the 1980s.

Gradually the Green Dot programme came into play in Germany which increased the volume of waste being collected. When the community collection programmes came on stream more widely we found that the amount of waste being generated was more than could be physically treated or consumed in Europe. Europeans – Germany in particular as it was ahead of everyone else – had the problem of finding buyers for their end product so exports to Asia soon became a necessity and at the time the biggest markets were India, Thailand and

Taiwan followed by Korea and Indonesia. Post-1995 to early 2000 China came on stream and today China is the dominant market for exported fibre.

But when I began focusing on paper the volumes of material were still comparatively small. We did not start experiencing big tonnages of waste in Europe until 1983/84 but probably more towards the second half of the decade. After that the community collection programmes really began being introduced and because I was already in business and had some experience of working in Asia I was able to take advantage of the opportunity. But in terms of collecting community waste Germany was the leader; in places such as the UK those programmes had not even started – that happened from about 2000 onwards.

After Germany the next countries to promote community collection programmes were in Scandinavia and Holland, thereafter every country started adopting similar legislation and the control regime of enforcing more collection of waste, of reducing waste going to landfill and of pushing to increase recycling targets followed mostly post-2000.

At the time there was something of a north–south divide within Europe when it came to implementing recycling programmes. The northern states took the lead and the southern states joined in later. But although they may have been slow to introduce their own systems the southern states have all been able to adopt the technologies and know-how that have grown over the years. Today in 2014 they are all on a par in terms of enacting collection regimes. As for achieving higher recycling targets and recycling grades Germany and Scandinavia are certainly ahead with other countries still trailing behind trying to reach the 70 per cent rate.

The European Commission's statistical office, EuroStat, released MSW recycling and composting rates for the 27 member countries in 2007 (see Table 6).

I don't know whether or not there is a cultural difference influencing the implementation of recycling programmes but in the last 15 years throughout the world there has been an appreciation and understanding that waste material has to be recycled. That concept has been accepted now and as investments have come in every country can introduce the very latest technologies and practices; Spain for example has a unique plastics collection programme.

It is noticeable however that there is a movement in Europe, particularly in the southern states, to opt for waste-to-energy (WTE) solutions rather than considering the recycling route. With energy costs rising and renewable energies and other sources of energy being looked at more closely, there is a tendency to divert recyclables at an earlier stage in the recycling chain. We are diverting tonnages to WTE without first exploring and exhausting all the other possibilities to recycle what we are throwing away. Conversely, in Germany and other northern states which have reached the maximum recovery percentages,

Table 6 MSW recycling and composting rates for the 27 EU member states, 2007.
(Source: Figures courtesy of EuroStat)

Municipal waste 2007			Municipal waste treated	
	Municipal waste generated, kg per person	Landfilled (%)	Recycled and composted (%)	Incinerated (%)
EU27	*522*	*42*	*39*	*20*
Austria	597	13	59	28
Belgium	49	4	62	34
Bulgaria	468	100	0	0
Cyprus	754	87	13	0
Czech Republic	294	84	3	13
Denmark	801	5	41	53
Estonia	536	64	36	0
Finland	507	53	36	12
France	541	34	30	36
Germany	564	1	64	35
Greece	488	84	16	0
Hungary	456	77	14	9
Ireland	786	64	36	0
Italy	550	46	44	11
Latvia	377	86	14	0
Lithuania	400	96	3	0
Luxembourg	694	25	28	47
Malta	652	93	7	0
Netherlands	630	3	60	38

Table 6 Cont.

Municipal waste 2007		Municipal waste treated		
	Municipal waste generated, kg per person	Landfilled (%)	Recycled and composted (%)	Incinerated (%)
Poland	322	90	10	0
Portugal	472	63	18	19
Romania	379	99	1	0
Slovakia	309	82	7	11
Slovenia	441	66	34	0
Spain	588	60	30	10
Sweden	518	4	49	47
United Kingdom	572	57	34	9

they are struggling to achieve even higher rates and so it is understandable that they too are turning to WTE, but it was not their first option.

The reason that the southern states prefer WTE is simple – market forces. As more investments are going towards WTE plants and with the cost of fossil fuels increasing people find that such plants are more sustainable and these alternative sources of energy are more acceptable by the market. As more and more investment is flowing in that direction companies operating WTE facilities are probably paying a higher price for the residual waste material which 'traditional' recycling plants cannot match. Secondly there is the issue of quality. If producers are at risk of having their products rejected either by a European or Asian mill, instead of taking those risks or installing more technology to improve quality they take the easy option of diverting to WTE. It is an obvious incentive.

Sadly, investment for setting up MRFs has not come as rapidly as one would have expected and now we are seeing a race to set up WTE plants in order to cope with the expanding waste collections. Sorting systems are not always able to produce the best quality and instead of sorting residual waste and extracting the recyclable material, because of these market forces, there is a tendency to use this residual waste as a feedstock for WTE plants as an easier option. The question for the near and longer term is: where is the increasing volume of waste that we are collecting and indeed anticipating going to end up downstream? Is it going to end up as recyclable material or as fuel for a WTE plant? Market forces will determine which system prevails.

Collection of waste is certainly going to increase along with population growth, as we have already discussed, but as we have also been saying there will be an equal increase in demand for more finished goods, more paper and cans for soft drinks. I find it difficult to support the diversion of this tonnage to WTE facilities on the grounds that we have plenty of waste and that we will have enough left to service the markets. There does not seem to have been any research on how this global demand will change over the next 25 years. The demand for fibre and plastic is only going to increase and if we burn it today we are depriving the market which will struggle to meet the growing requirement for raw material – both virgin and recycled. The logical consequence of running short of material means we will not be able to supply the market and therefore we will see an inevitable increase in prices and inflation. And what do we say to the consumers who participated in the recycling chain at the start of the process only for the result of their efforts to be diverted into WTE plants further down the line? Thank you for your contribution, for carefully separating your household waste but we have decided to burn it all.

As an aside we should not complain about the quality issue or even use it as an excuse for avoiding harder recycling alternatives. In the total mix of paper collections that we export about 50 per cent comes from commercial collections which are well within accepted quality standards. In the remaining 50 per cent that comes from community streams here again more than half of that will be of reasonably good quality. It is with the residue that we are struggling, trying to sort it into different quality standards. Nevertheless when measured across Europe, this is quite a big volume which is not meeting the stringent quality demands of the buyer. But China's Green Fence policy has done us all a favour because it has brought home the message that the reprocessor who is using this recycled material can only use it effectively if it is uncontaminated and made up to approved specifications. As this awareness increases across the world thanks to China, then this problem will cease to be endemic in the future.

While Europe may feel it is doing everything possible to recycle its waste there is no room for complacency. Despite numerous directives and well-meaning initiatives it is clear that Europe on the whole has a long way to go. Friends of the Earth reported in February 2013 that European countries only managed to recycle 25 per cent of their municipal waste instead of the 50 per cent called for by the EU 2008 Waste Framework Directive. When or if Turkey and Croatia join the Union the statistics will not look much better. These countries landfill 99 per cent and 96 per cent of their waste, respectively. 'Recycling targets are a good start', said Ariadna Rodrigo, resource use campaigner at Friends of the Earth Europe, 'but reusing products and materials and preventing waste in the first place won't be the norm until we have EU targets for these too'.

In fact Turkey is a big consumer of waste – one of the biggest consumers of steel scrap in Europe – and as waste collections in Europe increase we should remember that other countries where the middle classes are increasing will also be demanding this recyclable material. China started at one million tonnes and is now importing 30 million tonnes of fibre. Putting all waste together, China is probably importing just under 50 million tonnes, but that is not to say that China has reached its peak; it continues to grow. Vietnam and Myanmar are coming on stream as are other Asian countries. All of them have been deprived of books and newspapers, and as education expands, they will need more and more paper so the demand will increase for those precious raw materials.

Instead of talking about a small recycling loop within Europe it is the bigger loop around the planet where we need to be thinking. Are we Europeans thinking exclusively about Europe while ignoring the future needs of the African, Asian and South American markets or are we thinking about the whole planet? Europe as a major consumer is certainly generating and collecting more waste per capita, but will we continue to be the source of feedstock for emerging economies so they can produce the finished product which their markets will demand? Take a simple product such as tissue paper; once the emerging markets develop what will be the demand for this product in the next 10 or 20 years in Africa and Asia? I cannot even contemplate the scale of the requirement once those two continents combined are flourishing at their fullest capacity. The demand is certainly going to grow but whether we can use everything we collect ourselves is another question. However we must not forget that what we collect is usable raw material for other people in emerging markets and if we deprive them of it they will simply increase consumption of virgin resources. As these countries buy bauxite to make aluminium cans this will inevitably push up the price of our soft drinks because the price of aluminium will rise. At the moment our continued effort to export our surplus recyclable cans and paper to emerging countries allows them to make enough paper and cans to supply their local market demand. This becomes a virtuous circle.

I lay great store by meeting face to face and listening to the people at the sharp end – the mill owners, the reprocessors and the buyers. The big challenge that the whole industry is talking about is that the increased waste needs to find a home. We also need to have the right amount of value in the product for us to handle and process it and then get a fair price. But the tendering process is pushing the recycling companies to make higher and higher bids to get control of the tonnage and it is easy to see why. Every recycling plant has to run with the minimum furnish (feedstock) for which it was set up; if a plant requires 150 tonnes per day it must receive 150 tonnes, if it can only get 100 tonnes it is operating at two-thirds capacity and may end up having to close down. This is today's problem for many plants.

Operators are obviously feeling that they are having to bid higher and higher for the material, so high in fact that the price at which they have to deliver the material to the paper, steel or aluminium plants leaves them with little room to manoeuvre. While they accept that they have to deliver a certain quality, they feel that the pricing market and the volatility in prices mean that they are unable to deliver this quality. This is becoming a real problem: can they sustain the whole process and deliver a product at the right price or go half-way and produce a product which is of poor quality which the buyers are now rejecting.

We are now in a transition period. Everybody appreciates that the quality has to be right but the volume also has to be adequate to make sure the plants run profitably and are sustainable. The input price is governed by the output price of the end product they generate, the output price is in turn governed by the price of the virgin raw material and based on that price manufacturers can only pay a certain amount for the recycled material they are buying. All the recycling plants are running in a circle – quality/price/delivery/quality/price/delivery – and they are trying to do their best.

The need to tackle these problems of quality, price and volume is understood by all local authorities and councils across Europe and the rest of the developed world. Over the years local authorities have seen the bidding process evolve, not entirely of their own making. More and more companies have been building ever bigger MRFs which need greater supplies of material, perhaps 2,000–3,000 tonnes a week, and there are many plants of that size in Europe. The feedstock has been controlled by the local authorities in the community collection programmes and the companies have been making higher and higher bids to secure those contracts to feed their huge plants; the system has evolved into a price-driven auction. We are not objecting to that but once the highest bidder wins the contract the local authorities should become a 'partner' with that company to ensure that the product the recycler is producing is of the right quality and the right type. If there is a need for any improvement in equipment or logistical support some of that price should be clawed back to help improve these sorting and recycling systems. At the moment the highest bidder wins and the councils do not want to know what happens to the waste. This is a global issue but I recognise that the possibility of establishing some sort of partnership arrangement between contractor and local municipality is fraught with difficulty. Nevertheless some such arrangement should be considered because it is not a 'them and us' situation: councils need recycling companies to collect and treat the waste stream which they in turn are obliged by law to tackle. In reality everyone is involved in the loop; it is a never-ending process which is common throughout the world.

10

US market

The global waste market can be loosely divided between the developed Europe, North America, Japan, Australia and New Zealand markets; Latin America/the Caribbean; and the emerging economies of Africa, the Middle East and Asia. The USA to all intents and purposes is the market leader in recycling with the solid waste industry estimated to be worth about $55 billion in 2011.

This has happened out of necessity – the size of the country, industrial development, the population and the volume of waste. It is the world's largest single national economy with a nominal GDP of c.$16.6 trillion in June 2013. However, if the rest of the world tends to follow the lead of the US in terms of waste management, that could be a concern.

Unfortunately there are different schools of thought. At one extreme some argue that waste does not need to be treated and can go into landfill because America has plenty of space to take care of all it discards. But the point is that many of the goods Americans consume are imported and the countries producing and supplying America need raw material to manufacture those products. In their own region, principally Asia, there are not enough raw materials available therefore if America did not recycle and promote recycling, which fortunately it is doing, then we all would lose the benefit of that material for good and the recipient countries would be starved of their essential feedstock.

At the other extreme there is the approach we have already mentioned of putting everything into one large bin and allowing the dirty MRFs to sort it out or sending it all to WTE plants. As always with extremes there is a grain of truth to be extracted from each, but there is a much larger middle ground where most of the solutions lie.

If America 'switches off' the tap supplying recyclables what options does the manufacturing world have? We can look for recycling material from other regions but the volumes are simply not there. America produces the lion's share of the 50 million tonnes of waste paper, plastic and metal being traded around the world. So without the US input we would need to source our raw material from natural resources and we know the consequences of that approach. Recycling is important and fortunately the USA is very supportive. A lot of recycling material exported by America comprises commercial tonnage but also some community waste; America will remain for the foreseeable future one of the major suppliers of recyclables unless there is some radical swing in policy.

For all the new technology being developed, collection systems are the key but they vary wildly from country to country and in America from state to state, even from city to city. Many are driven by cost rather than just environmental issues and rightly so because the commercial aspect of waste management plays an important part. Because the volume is so high, the challenge for America is to find a compromise between single bin systems so they can tackle their waste stream in the most cost-effective way and finding a way to preserve the maximum possible waste to recover and reuse in order to extend the lifecycle of every piece of waste that ends up in the bin.

Europe is driven by tough legislation whereas in North America there is probably nothing as severe as the waste directives issued from Brussels. This is not to question the environmental credentials of the USA but they are faced with the pressures of vast consumption in the major cities and a large population, albeit one which can be spread thinly over huge distances. In some situations it simply makes no sense to build an elaborate MRF or a WTE facility. Nevertheless, natural resources are getting scarcer and rather than focusing so much on climate change we should be looking at how we can save tonnes of recycling material in regions where it may not be cost effective to collect it in order to maintain a steady supply of raw material for industries globally.

Waste management has grown and developed tremendously since I started trading with the USA. Then there were purely commercial streams, computer printouts and some cardboard and waste from printers. There were only a few grades of paper making the rounds whereas today there are many more streams and qualities. There is more money in the industry and economics as always is the driver. Stakeholders that have come into the business have enormous financial resources and see waste management as a profitable sector as well as being a socially and morally acceptable area on which to focus.

. A study commissioned by the Environmental Research and Education Foundation in Washington, DC in 1999 found that the solid waste industry in the US generated total revenues of $43.3 billion:

> the private sector generated about three-fourths ($33 billion) of this amount. The publicly traded companies accounted for 47 percent ($20.6 billion) of the total, while privately held companies represented 29

percent ($12.4 billion). The public sector accounted for 24 percent ($10.3 billion) of the total revenues.[71]

All these figures have now grown. The *Environmental Business Journal* explained the drivers at work: 'Economics, public opinion, and government mandates will increasingly demand that more value is recovered from our waste materials', said George Stubbs, senior editor. 'Municipalities are moving forward with diversion programs – in many cases motivated by state mandates. And leading solid waste companies will continue the process of re-branding themselves as businesses that also extract value from the materials they collect.'

According to the US Environmental Protection Agency (EPA) in 2010 Americans generated about 250 million tonnes of trash and recycled and composted over 85 million tonnes of this material, the equivalent to a 34.1 per cent recycling rate.

With so much value at stake we have to ask how we can satisfy the balance between the right type of plant, collection streams and the economics of energy. Business is playing its part in two ways: it is helping to keep the planet safe by considering all the options not just the easiest one; and it is fulfilling the role of a solid business by choosing to back solutions which make economic sense. America has to be the leader in this enterprise but I sound a note of caution particularly during these times of financial austerity post-2008.

After recent talks that I have given in America I have noticed that people tend to focus on producing recyclables to match their operating systems. They ask me where there is a market for such and such a product, instead of analysing their waste input and producing a recyclable raw material that meets global demand and specification. Such markets exist but the product may need to be adjusted to meet requirements. Unfortunately, the recyclers are unable to make such changes. The facilities are only producing what their equipment allows whether they are restricted by know-how, understanding or technology. The challenge for companies setting up new facilities is to allow for change. It may take four or five years to build a new MRF during which time demand will have changed – it always does. So rather than be fixated on the system itself the company commissioning the build should keep a closer eye on the market and the feedstock it expects to use because it might alter in weight, size or consistency in the relatively short time it takes to build the new facility. Has the architect allowed for expansion, a new sorting line or a new development which has to be incorporated because the demand for the recyclates the company intended to produce has evaporated?

Building a standard MRF was acceptable in the 1980s and 1990s when the buyer of the material was driven by the costs, which at the time were low. The

---

71 Waste360.com

low cost meant that time could be spent re-sorting the waste to produce a usable raw material to feed the machines, but as time has gone by with tendering costs, increased cost of collection, logistics, increased investments in bigger MRFs and sorting systems, so the cost from the point where the waste was generated to where it becomes a raw material has increased substantially. Nowadays the mill buyer is rethinking. A grade of recyclable that may have cost $25 in 1990 may now cost $100 plus. With these increasing costs it is not possible to re-sort at the mill end therefore the buyers are increasingly demanding that the product meets their requirements. But of course it may be too late because the recycling companies have already invested in a sorting system meeting the demands of the collection streams running at the time the facility was commissioned and no allowance in many cases has been made for changing trends or even room for expansion.

That is the transition we are going through at the moment and that is why it is the consumer of the recyclable material that is the most important part of the chain. Those systems and those countries that are willing to understand and talk to the consumers and create a product that suits the consumers are certainly going to make a lot of progress in meeting recycling targets. In addition investors and industry are cautious about where to put their money: should it be in a super MRF, should it be in mass-burn (incinerators) or some form of anaerobic digestion (where America actually lags behind Europe).

Let me give a specific example: that of e-waste, the fastest growing new waste stream. According to the United Nations, America created 9.4 million tonnes in 2012, the most by any country. Only seven years earlier it produced 2.2 million tonnes which marks a significant jump. China came second with 7.2 million tonnes but when size of population was taken into account the US generated 30 tonnes per head compared with China's 5.4 tonnes.

Managing e-waste is a growing science and yet with tyres, an age-old problem, people are still looking at different ways of extracting oil or separating out the rubber. So there will be change. Until now it has been the consultants, architects and advisers who have been designing the plants without a close eye on the waste streams, and in the past whatever we produced the buyers accepted because the labour costs were lower, the choices were fewer and the prices were low. Today all this has changed. Prices have increased across the board and every manufacturer globally is exploring different ways of increasing their recycling capability which means we have had more recyclables available. Buyers choose the quality that is closest to their requirements and the waste that is nearest to their plants to cut the delivery time. As a result over the last ten years these aspects have had a more powerful influence on the business and even some relatively modern facilities have quickly fallen out of date and been forced to close.

Plants dating back just to the 1980s and 1990s will probably sustain themselves producing a locally acceptable quality servicing the needs of local industries but as the market opens up and the choice widens even for the local consumers, every manufacturer naturally will go for the best quality available. Ultimately every industry, consumer and user of the waste will be governed more and more by costing factors and will want the best available feedstock to manufacture a higher quality end product.

In the long term, perhaps ten years from now, costs and breakeven points will come more into focus. If the cost of alternative sources of raw material to natural sources starts going up because the supply and availability of virgin material is falling, then at that point we may have to reconsider if it is wiser to invest more money to segregate the waste to get a better price rather than to burn it for energy purposes. Energy may appear on paper today to be a better option in some cases but later on as the availability of natural resources tightens and raw materials are in short supply the prices of both recycling as well as natural resources will go up. Perhaps then WTE may become the very last option. Ten years – the mid-2020s – is not that far away when we are planning future generations of MRFs and sources of alternative energy. As I have stated earlier, where we have comparatively modest amounts of waste, we know we are short of modern MRFs, but what should we build, where should we build them and how big? Planning permission alone is a costly and slow process. As the winter approached in December 2013 the British Government revealed that every single county in the country bar Cornwall was a potential site to explore for shale gas using the controversial fracking technique – the planning and the protests will be endless both for shale and MRFs.

What advice can one offer? There are certain regions of the world such as America, Europe and Australia where there is an abundance of raw material and plenty of people who are ready to use recyclates as an input for their industries, but in emerging economies the volume of recycling material can be more than available industries are willing to use. There is room for building many industries across Africa but consumption is a continual process, so the waste is certainly being generated albeit at a slower rate than Western economies. If the waste is being generated what is happening to it? Is it going to landfill or just into great waste dumps on the outskirts of the cities?

Commercial waste streams in the major developed cities are already supersaturated, so where does a recycler get their waste? Councils are looking for the highest bidder, but having paid the highest price what does the processor do with it? They have to process it into a quality that has a market but when they have met all their costs are they then able to sell their product at a price where they are making some profit and also supplying the buyer with a raw material that is going to increase their yield and revenue?

The only sound advice regardless of whether we are talking about America or an emerging nation has to be to focus on the right technology based resolutely on the waste stream to ensure we extend the lifecycle of that waste. This also means that we should respect the way it is collected because only then can it be used as a raw material, indeed only then will it be acceptable to the buyers. America is certainly ahead of Europe in building new systems but there are many different technologies available.

Waste management as a subject has grown tremendously in the last two decades and alongside waste management WTE is making similar progress. The people who support the WTE school of thought are investing a great deal of money into systems which are cost effective and financially viable in order to use the waste and produce either a recyclable product or some other form of consumable. For example in the last ten years I have seen the growth of sawdust mixed with plastic waste to produce a form of composite wood with a reasonably long lifecycle which is being widely used in America for decking purposes. In the past sawdust might have been used only as animal bedding or even just burned with little thought being given to creating a commercial product.

We are certainly going through a revolutionary phase with many new technologies being tried and tested. But the success of any of these technologies depends on how closely situated the MRF is to the waste collector and also whether there is an adequate supply of waste, because generation of waste depends on the population within the region being served. Many prototypes are successful but fail to deliver when translated to a commercial scale sometimes through lack of feedstock.

Western economies led by America have raced ahead while emerging economies are in a state of flux, finding energy the main driver; booming economies such as India where domestic energy is in short supply find fuel costs very high. Japanese citizens are questioning the whole future of nuclear power following the Fukushima Daiichi disaster on 11 March 2011 when the plant was hit by a tsunami. How will other countries look at nuclear power now? Inevitably they will find WTE the cheapest and fastest solution and not really be interested in looking at waste to see how they can recycle, recover and reuse it for the future. At the moment the priority is economics – if they can burn waste for energy in a cheaper and safer way by using perhaps a mini WTE plant they will probably go for that option.

When the US waste industry talks about the next generation of waste management solutions it seems there is only one way forward. According to the Solid Waste Association of America (SWANA) about 80 communities have waste-to-energy projects which some regarded as a slow uptake – they want more WTE plants.

11

Latin America and the Caribbean

People speak of the Global North and the Global South when looking at the Americas. Latin America and the Caribbean is the Global South, a wide region representing many countries with different levels of modernisation but with common issues as regards the recycling of waste and its challenges. It is rapidly growing in terms of GDP, consumption and consequently waste creation. What the countries have in common is a lack of resources, a lack of know-how and/ or the desire to 'get rid of the problem' in the easiest and most cost effective manner, which too often means dumping in unregulated sites, landfilling or incineration. The key challenge in the face of this economic expansion is not to lose sight of recycling while the other sectors are growing. We are not making a case here for less production, rather a coordinated strategy which addresses the impact of growth on a community, including its health.

With that agenda we must again start from the basics. We first need to look at how waste is collected, not just in the commercial sector including industry, supermarkets and printers and what happens to it thereafter, but also at community level. Many of the great cities of the region have been growing so fast that it may have been too easy to neglect the impact of that growth: Greater Mexico City has a population of *c.*20 million, Buenos Aires 13 million and Rio de Janeiro ten million.

Waste recycling is not given the same priority as the economy, health, even crime and yet it is precisely as the economies grow that the waste problem escalates and as it is neglected sanitation becomes a serious problem. As for crime there are areas in some of these countries where it would be too dangerous for normal waste collection services to operate without protection against a strong gangland culture.

Secondly, and despite the obvious challenges, we must start working on regulatory drivers to promote domestic collection streams – this is easier said than done in some of the poorest neighbourhoods but unless robust frameworks are established collection will be erratic at best.

Thirdly, this region is very slow in building up the export drive for the waste they are undoubtedly able to generate. There is potential wealth in the waste that is so readily discarded but it is a potential which can only be grasped effectively with detailed knowledge of the wider market place. As the essential elements of a waste collection and treatment programme are developed at government level the authorities should look to the opportunities that lie beyond their borders because recycling is a multinational business. Latin America and the Caribbean needs to focus on how it can promote more recycling and become a global trade partner. The World Bank predicted that the region's municipal solid waste would increase from 131 million tonnes in 2005 to some 179 million tonnes by 2030, most of that in the major cities. The only certainty is that waste will not decrease and the region needs to be able to respond to the constantly changing demands of the buyers.

I remember speaking at a conference in São Paolo, Brazil on the topic: Is there a global market for recycling and how can Brazil be part of it? Following my presentation it was clear that the delegates were keen to join this global supply chain but the basics of supply (waste from collections) and infrastructure (recycling facilities) were not in place simply because they had not been a priority. The only significant waste collection in the country at the time was supported only by commercial streams. After my presentation I visited some metal scrap collectors and it was apparent that they certainly had more material than they could use but at the time there was an internal embargo in Latin America to contend with – tonnage could be moved within that region but not outside. They were not getting the right price for their waste. This is an example of coordinating international regulatory controls. What is important is that we ensure that the economic side of this business is allowed to grow on a global scale because one of the drivers of recycling is economics and the people collecting the waste must be able to get the best price per tonne so they are able to produce a product which is of high quality, less contaminated and better sorted. In a region as significant as Latin America and the Caribbean steps should be taken for countries to work together.

It was also obvious to me during my visits that the local industry was unaware of global challenges in the international market: the control regimes in different regions of the world, what the requirements are, what grades are in most demand and how they can make the grades better suited for export. I visited plants for metal and paper and it was illuminating for them when I was able to explain how their market could grow given a better perspective of world trade.

By 2014 the world's focus was increasingly being drawn to South America as the temperature of the economic tiger economies in the Far East began to cool and attention shifted in that direction drawn even perhaps by the World Cup in Brazil. But that focus has also put the spotlight on what needs to be improved – the squalor of the *favelas* and the shanty towns. And yet this is where we must find part of the solution. The '*recicladores*' – just one of the names given to the vast army representing the informal sector of waste collectors in these cities – should be included because it is estimated that they are responsible for up to 90 per cent of the recyclables recovered from the waste stream.[72] This is no ragtag army. The *Red Latinoamericana de Recicladores*, or Latin American Waste Pickers Network, is made up from groups of workers in 15 countries in Latin America and they held their First World Conference of Waste Pickers in 2008. According to one estimate nearly a quarter of a million Brazilians alone are involved in waste picking.[73] Indeed throughout the region, in Ecuador, Chile, Paraguay and Bolivia, the waste pickers are regarded as an integral part of the recycling solution and often work closely with the formal sector devising strategies and action plans. As Latin America is one of the most urbanised regions of the world, this is where the work must begin.

Dr Alfonso Martinez, New Projects Director, Mundo Sustenable, wrote:

> By 2050, 90 per cent of Latin America's population will live in urban areas. This high rate of urbanization coupled with the global economic growth has resulted in a waste management crisis. Municipalities find themselves unable to keep up with providing services and infrastructure to the urban populations.[74]

There are always two streams – the commercial stream and the community stream. The waste from the community stream is put in bags and thrown away on to waste dumps and these community streams have certainly created employment, albeit unhealthy work, for people to pick over and find items which can be resold. It would be preferable not to lose the service of those people but improve on collection streams so they can be more gainfully employed in a healthier environment to continue what they are doing for their own survival and probably resulting in a much higher percentage of recovery than they can generate on the current informal basis. Recycling in this environment is a labour intensive process and I know there are moves to formalise the *recicladores* into cooperatives to give some kind of structure to the collection process, to train the collectors to sort organic and non-organic items and in that very basic way improve the quality of the waste stream. The temptation is always to go for a 'modern', high-tech solution

---

72   Columbia University – *Advancing Sustainable Waste Management*
73   WIEGO Fact Sheet: *Waste Pickers Brazil* – Helena Maria, Tarchi Crivellari, Sonia Dias and André de Souza
74   Columbia University – *Advancing Sustainable Waste Management*

to solve the waste problem but I do not support this approach at present because we are ignoring the vast potential of the informal workers in recycling which in some cities has resulted in a higher rate of recovery and reuse than in the West. For example before the collapse of law and order in Cairo in 2013 the recycling rate was 60 per cent compared with Rotterdam's 30 per cent.[75]

So my first message would be to get the authorities thinking objectively that for every kilo of material that is not collected and processed correctly, sold or reused in some form, they are losing revenue. This is an opportunity for industry to become more proactive. They have to collect and reprocess more waste to produce raw material for domestic or export usage. Obviously what is not processed or landfilled is certainly a lost economic opportunity.

Secondly, they must understand what is available in terms of technology and adapt it to the local collection streams; in other words what is collected locally must have a specially adapted technology that works in those circumstances. Above all this does not mean investing in the very latest systems. This is true for every country, particularly the emerging nations.

Thirdly, they must take advice to understand consumption in other economies and see what is consumed where, what the demand cycles are away from their market and how Latin America and the Caribbean can meet those demands. This is critical for all those emerging nations seeking to become part of the international recycling loop but it requires knowledge of the global recycling market place.

Of course this all demands significant levels of investment and there is often an awkward division between the municipalities responsible for waste collection and the governments which control the purse strings. We need to encourage the finance houses and investors from regions within Latin America or even outside to consider working to build the industry because it has a potential for major growth. Already quite a few European companies have moved offices to Mexico and Brazil in various sectors but more needs to be done. This is where governments and banks in these countries should invite investors to their region and promote greater generation of recycling material.

But it is also the duty of governments to educate both the community and industry. There needs to be a new emphasis on having more environmentally friendly factories to ensure that the waste water, chemical waste or semi-solid waste is properly treated before it is disposed of. Effluent waste water must not be allowed to pollute rivers and there need to be better waste plants to detoxify waste from industries before it is released for disposal.

The Caribbean is of course a tourist haven and the quality of the water around the islands is vital to attract the visitors and also for its thriving fishing

---

75   UN Habitat 2010

industry. The consequences of untreated waste water being released or even the unsightly appearance of waste dumps are incalculable. At a conference in St Eustatius, Hugh Riley, the Secretary General of the Barbados-based Caribbean Tourism Organisation, said:

> We must practise sustainable water-use policies and observe appropriate waste-water management practices. We must not only enact, but also enforce legislation that regulates the proper disposal of waste in the waters that wash our shores, and we must severely punish all violators, because they endanger our health and jeopardise our children's future.[76]

Latin America is certainly exporting metals but in the lower tiers of plastic and paper we do not see them as a major partner and therefore the worrying question is: are we letting all this waste go to landfill and dump sites or is it all being consumed in the domestic economies? My fear is that it is not all being consumed by the domestic economies because we are not collecting everything that is possible which means potential waste recovery and therefore potential profits are being lost. Brazil, Chile, Argentina and Mexico are all leading players in the region and they can do more in promoting recycling programmes for the community and probably ensuring that the industrial sector is able to utilise more of the waste that it generates.

The problem is that recycling is not seen as a priority for many emerging economies but governments do have a duty of care. That means introducing regulations to force local municipalities to have waste collection programmes which would ensure that the increasing volumes of waste are collected and that the recovery methods are friendly in terms of ensuring that the recyclables produced are suitable for the end user/buyer. The state of Mato Grosso in Brazil was the first Latin American state to pass a specific law to prevent e-waste. Part of the problem is that in some countries even major states and provinces do not have the power to introduce environmental legislation which can lead to inefficiencies even where there is a desire to act.

This is a golden opportunity for the region but there are issues – social issues and economic issues. In the run up to the 2014 World Cup, Brazil has seen demonstrations with protestors challenging the expenditure on such sporting events when they felt there were more pressing social requirements. One cannot impose the waste recycling solutions which may be appropriate to New York on Mexico City and one has to work with the realities on the ground; recycling trucks cannot operate in the chaos of the narrow slum streets and in any case there is the livelihood of the countless thousands picking over the dumps to consider.

---

76 *Jamaica Observer*, 26 September 2013

One also has to be realistic about the budget priorities of Latin American and Caribbean countries although it should be said that many of the Caribbean islands including Jamaica, Antigua and Barbuda, Dominica, Grenada, St Kitts and Nevis, St Lucia and St Vincent and the Grenadines have made conscious efforts to attract outside investment and launch initiatives to tackle the waste problem and establish sanitary landfill sites. All these islands want their economies to prosper and naturally devote maximum financial resources in that direction but the logical consequence of a booming economy is a commensurate increase in waste.

As ever the most effective steps depend on education and funding. Education includes raising the awareness of municipal authorities about what can be done and what they should be doing and constant reminders to the general population of why recycling matters and how it can benefit their livelihoods and even their general health.

The challenges the region faces are sadly common to a greater or lesser extent in many parts of the world and while tougher controls on what happens to waste are called for, this will inevitably create opportunities for criminals to try to get round the system. Even officialdom finds it hard to cope. Greece, which assumed the rotating Presidency of the EU in 2014, was facing court action for the many illegal waste dumps around its islands; it may be in economic difficulty but the waste keeps coming just as it does in emerging nations which still lack many modern facilities. For every law that is enacted someone will find a way of getting round it, apparently uncaring about the damage carelessly dumped toxic waste might be doing to the environment even when it is the beauty of the environment in places such as Greece or the Caribbean which attracts the tourists and builds up business in the first place.

Trash, waste, rubbish is at the bottom of every municipal agenda and yet it is a constant which is ignored at our peril. We don't want to treat it anywhere near our homes and we don't want any restrictions on what we throw away. So long as there is a great enough distance between us and the landfills we are happy to forget about it and too often that distance means: let's just dump it in Africa.

12

Africa, Middle East and Asia

Collectively, while some countries in Africa, the Middle East and Asia may have been left behind when it comes to recycling technology, these regions hold an important key to the future of global waste management which is being largely ignored. Or, to continue the metaphor, the lock is being forced and unless we take great care we may never be able to open the door to the potential each of these regions holds for us all.

Waste collection in these regions ranges from the virtually non-existent to moderately advanced; if programmes have been established to any degree they tend to be for commercial not community waste. They are at a developmental stage of growth and the levels inevitably vary considerably. Even in some of the more advanced of these nations there is often no coordinated legislative structure and investment programme driving the business of waste management. In 2014 it is hard to imagine that any collection at all is happening in war-torn countries such as Syria, South Sudan or the Central African Republic while in the blossoming cities of China waste generation is being handled leaving the countryside to fend for itself.

China has been the world's largest generator of waste since 2004, overtaking the United States. It is estimated that it will probably be producing twice as much municipal solid waste as America by 2030.[77] Malaysia on the other hand has begun to introduce regulations requiring community collection programmes while Thailand and Indonesia have nothing at a significant level.

Again there is the impact of urbanisation to consider – the pull of the cities on workers and their families from the countryside in search of employment. The World Bank estimates[78] that the Asian urban population will reach 50 per cent of the total population by 2025. The urban sprawl will simply continue but the unstructured growth does not lend itself to organised municipal services

---

77 *What a Waste: A Global Review of Solid Waste Management* (World Bank)
78 World Bank, 2003

such as waste collection and invariably the waste – much of it organic – finds its way on to the nearest dumpsite, not even landfill.

Throughout the 1970s, 1980s and 1990s we in the West have benefited from the poorest nations of Africa not only by exploiting their minerals but by using them as the developed world's waste dump. Anything we didn't want and thought too hazardous to handle or treat ourselves we happily shipped off to the continent, turning a blind eye to the perils facing the generations who picked over our polluted presents. In 1987 the Basel Convention first got people to acknowledge the implications of what we were shipping round the world and tried to impose regulations to bring this illegal cross-border trade to an end. More initiatives and more conferences followed. The Organization of African Unity sought to ban the import of hazardous waste but a combination of greed and desperation for some form of income ensures that the traffic continues to flow.

Africa has enough problems of its own without importing any from the West. The population of Africa will more than double to 2.4 billion over the next 40 years according to the Population Reference Bureau of America with African mothers giving birth to an average of 5.2 children compared with the European average of 1.6, thanks in large part to better healthcare.[79] The International Monetary Fund predicts economic growth in sub-Saharan Africa to increase to 6.1 per cent in 2014/15 bringing in new investors. So more Africans will achieve a better standard of living, will produce more, will demand more and of course dispose of greater quantities of waste. As a comparison the OECD group of 34 industrialised countries generate c.1.6 million tonnes of MSW per day while sub-Saharan Africa produces around 200,000 tonnes a day.

Reinhold Schmidt, President of the BIR Paper Division, sees an urgent need to help developing countries:

> It is not enough for recycling to be limited to only a few countries. There is only this one world and we should all participate in the conservation of our natural resources. Where countries in which recycling has not yet started or only takes place on a small scale, the rest of the world should help create recycling opportunities. To export our waste and dump it in these countries should be prohibited as soon as possible.

The Arab world is seeing many of its cities expanding at an unprecedented rate with growing populations, enormous infrastructure projects and booming economies. With that growth come social and environmental challenges and there have been significant efforts to meet those challenges. In recent years we have seen quite dramatic growth in terms of collection, sorting and trading of

---

79 *Daily Telegraph* 13 September 2013

fibre, plastic and aluminium. Saudi Arabia is one of the leading MENA (Middle East and North Africa) countries with very strong and robust collection programmes for commercial waste, excellent recycling facilities and also paper plants using waste as a feedstock for running the machines. New facilities which are dependent on waste paper are being installed all the time in Saudi Arabia as well as other countries in the region. What needs to grow simultaneously are strong community collection programmes so we can reduce what goes to landfill, increase collection volume and ensure that the waste that they cannot use can find an alternative home somewhere else around the globe to become part of our closed loop system.

> Landfill mining of desert dumps has increased as cities expand. Saudi Arabia has been witnessing rapid industrialization, high population growth rate and fast urbanization which have resulted in increased levels of pollution and waste. Solid waste management is becoming a big challenge for the government and local bodies with each passing day. With population of around 29 million, Saudi Arabia generates more than 15 million tons of solid waste per year. The per capita waste generation is estimated at 1.5 to 1.8 kg per person per day (EcoMENA consultancy).[80]

By contrast Qatar's official website says it is generating 28,000 tonnes of solid waste per day of which a large part is landfilled. But Qatar recognises the potential of a greener environment and recycling by saying it could enjoy revenues of up to US$663 million if the waste streams were effectively treated. It, along with other countries in the region, are creating centres of excellence to manage growing waste streams and capture the revenue and environmental potential of recycling.

So what do I mean, the locks are being forced in these countries? In their eagerness to sell the latest recycling technologies companies are flocking to these emerging economies, staging trade shows and dangling the bright, new sorting and treatment systems in front of potential buyers who are invariably government representatives. But are they really buying what they need? When we are talking about community collection programmes technology is superseding the basics. Firstly, it is very important for every country and every city to educate the citizens about how the waste should be segregated because if the primary segregation is not done correctly the whole idea of using the technology becomes irrelevant.

Secondly, there has to be some form of regulatory control or framework which drives the programme and then depending on the waste that is collected

---

80   http://www.ecomena.org

that should define the recycling programme needed for the area. Many countries appear to be pushing themselves to collect all the waste in one big heap, then using it as a feedstock to run WTE plants, which is detrimental to the planet because we are going to permanently lose the paper and plastic by burning it. This is not a particularly African problem; there are Western authorities who quite happily sign contracts handing all their waste over to WTE companies solving their waste target at a stroke. As readers will know by now I believe this is a short-term benefit which will prove costly in the long run.

So the emphasis should be on understanding the complete process: collecting the waste; educating the citizens or householders; the need for a regulatory framework to guide collection streams and show local authorities how waste should be segregated and collected; to create more sealed landfills so we are able to landfill the residue of waste in a way which protects the surrounding soil from toxicity. Only when the complete process is understood can we decide on the route to be followed and thereafter develop waste facilities and technology to fit what is going to come in. Too many countries, companies and investment houses are putting in plants to receive, segregate and bale the waste simply to feed these huge WTE plants.

Most of the countries in the region now embarking on waste programmes are racing to catch up with the rest of the world with the belief that once the technology is in place recycling will follow – this is a mistake. It is a case of taking action without first having a clear vision; ultimately this will lead to failure. It is only once the local education and understanding are in place that the technology should be selected to fit what is being collected.

At the moment we have material that is not collected in the right format, there is no education, no regulatory drive and no proper collection systems, so the pressure is on emerging countries to find an instant solution for what they should do with their waste. In Asian countries it is ending up on informal dumps and landfills. Because of the need to drive the tonnage away from landfill the easiest alternative is to have the waste compacted to feed WTE plants. And since energy is a big issue for many Asian countries and it also helps them to say they are building renewable sources of energy, WTE certainly wins the argument at the moment. But really what we are not doing is putting into force what we have spoken about earlier: education, regulatory drivers and collection systems so that we are able to collect the right quality of waste to recover and reuse it. We are certainly moving in the wrong direction here which will have an effect in the next 10–15 years when with increasing demand for fibre we will probably have burned much of what we need.

The exception to most of these rules in the Middle East is Israel. They recycle about a quarter of their waste and what they don't reuse goes into state-of-the-art landfills servicing most of the country through well-established municipal waste systems. Israel faces the common problems of a rapidly growing population and rapid commercial and industrial development. But it has brought in strict laws such as the Tire Recycling Law and packaging laws which put the burden of collecting and recycling packaging waste on manufacturers.

The Middle East, Africa and Asia are going to be important suppliers of raw material through their own waste streams; however by rushing into unsuitable MRFs or WTE plants we are leaving future generations exposed. I am a supporter of WTE but we should only be using the residual waste that cannot be sorted or is not fit for sorting as feedstock for these plants. Unfortunately the lobby against this thinking says there is no need for sorting. They say that we have got enough fibre to meet the demand, but that is not the case. We have to find how far down the chain in collection we should go so that we can recover as much paper, aluminium and plastic as possible before burning it. What these emerging countries are doing is taking the easy option by collecting and burning because it allows them to get rid of the mountains of waste, reduce landfills and create energy which is the pressing need of today. They are succumbing to the charms of the sales people promoting the latest technology.

Africa has a particular challenge. Its per capita consumption and generation is much lower than the other countries in the region, but although it has a huge and growing population per country the people are spread over vast areas and the volume of waste to be collected in particular towns and cities may be much smaller than that of, say, a booming metropolis in the Middle East or Asia. So it is the collectable volume of waste per square mile which is the big question and this is what makes burning or even dumping the easier option. Before a solution can be found we must recognise the problem and we need to look at major regions of these countries and then devise effective recycling programmes. One size does not fit all.

China has unique issues: not only is it the world's leading importer of so much of our waste but it is also the powerhouse, producing many of the products we eagerly consume in the Western world. China has rapidly expanding major cities which are industrial hubs in their own right and the consumption within cities spread over perhaps 50 miles will generate vast quantities of waste. Both commercial and community collection programmes have been growing steadily, commercial waste especially. In the last 10–15 years (to 2014) China has come a long way in recycling because its industries are geared to using recyclable material as a result of government and local authority work to promote recycling.

Such green awareness is of course to be applauded but the consequence for the West is that the Asian buyers, like all modern day industries, are acutely cost driven and constantly compare the cost of imported recyclables with their own domestic collections. They also look more closely at the machining and preparation time in terms of the finished product from a tonne of virgin material compared with recyclables; in other words measuring the yield factor that they generate. In addition there is the time it takes to ship material thousands of miles to its ports and the financial exposure in terms of various factors including currency values: renminbi versus US dollar, Korean won versus US dollar, Malaysian rupee versus US dollar – all this has to be calculated.

Furthermore the shipping lines are now using bigger and bigger vessels compared with 35 years ago when they were carrying a thousand 20-foot containers which used to be the normal vessel size. Today the larger ships operating on the high seas can carry 16,000 20-foot containers. However, the voyage time has increased over the years and now it takes 35–40 days to sail from the UK to China whereas in the past it might have taken 22–25 days to complete the same voyage. The Asian buyer is therefore constantly evaluating the cost of setting aside funds for that extended period. All these variables add up when they are calculating their yield factor. Why should buyers risk so much importing the recyclable tonnage which takes so long while virgin material is probably available on the doorstep for just-in-time delivery? Virgin material suppliers now hold stocks in these regions and they can deliver within a week of the buyer placing an order. With all these multiple factors coming into play it is also becoming more challenging for Europeans to export their recyclables to these markets.

As Western nations woo these emerging markets, the trade missions headed by the great and the good trying to secure lucrative contracts, it would seem that there should be more activity in trying to win control of the important raw material which makes up every transaction – each TV consists of different components and has to be packaged and wrapped for shipment and sale. The feedstock to make all these elements is vital and yet there is limited activity on the mergers and acquisitions (M&A) front.

Until 2014 one of the few companies making major inroads was Novelis, the world's leading producer of rolled aluminium. It is a company spun off from the Canadian-owned Alcan Inc. in 2005 and acquired two years later by Hindalco Industries of India which made Novelis part of the Aditya Birla Group. They have taken on a number of non-ferrous plants around the world supplying them with scrap and making them a fully integrated operation. Novelis say that the recycled content in its products has reached 43 per cent and it is now the world's largest recycler of aluminium in the world.

Another is the ArcelorMittal Group, the leading steel company, which has expanded into steel mills across the world and has also acquired steel scrap plants in order to have full integration of supply and sourcing of its own steel scrap over and above what it buys in the market to support the demand for feedstock.

However, in the other sectors such as paper and plastic it is surprising that other than the major European and American leaders there is not more activity in the M&A market. This is the real sector of opportunity in the 21st century. It is no use having big modern manufacturing facilities if there is no control of the raw materials/feedstock. In the next 20 years the most important actor will not be the industrial producer but the feedstock supplier – the raw material supplier – because as volumes of virgin material decrease we will see more volatility in the prices; dependence on recyclable materials will increase and those who control market share will be the market leaders. Backward integration in managing more and more of these recycling businesses would appear to be an obvious step, but there does not seem to be much interest in this sector from outside Europe and America.

While we are discussing M&A activity I am convinced there is an opportunity for these finance houses to help SMEs to take the next step in their own growth. Many of them have a wealth of experience built up sometimes over several generations. But if these firms are not given the help they need – perhaps even to expand into overseas ventures – they will simply disappear and that knowledge will be lost.

Bringing us back to Africa, for example Nigeria or Angola where sources of energy are plentiful and therefore cheap, I would advocate building up an industry to supply steel ingots by utilising the rusting cars and discarded shipping vessels and exporting the ingots to all the industries in need of material. The scrap metal would be put through a shredding machine to produce shredded steel which is a raw material for steel industries. Instead of just abandoning the wrecks they would become a readily usable raw material for all the engineering facilities which use steel as an input. Obviously the quality of steel which is oxidised from electric arc furnaces is different from the carbon steel from the big blast furnaces, but that is a secondary argument about the purity of steel. These regions could do very well and this is just one example of an industry waiting to blossom. Such an initiative would not only be environmentally sustainable but would also increase local employment.

For recycling to succeed for the end user who is going to use the recyclable material we have to make sure that value addition down the chain is done in an effective manner. Demand for finished goods is only going to increase as the population increases. To meet this demand we have only two choices: one using natural resources to produce the goods we require which unfortunately are in short supply; or two using the natural resources alongside recyclable raw material so we extend the lifespan of the natural resources. If, as is my hope, we are going to go for the second option, the successful recycling company is the one that can guarantee delivery of this material.

In the next 20 years the price of natural resources will increase substantially and the importance of recyclables will increase commensurately. Recyclables are collected from the community and commercial streams. For industry to be economically successful it has to ensure its raw material supply chain is well managed right down to the householder. So the companies that are going to have better control over getting the recyclables will be the market leaders while those who are just depending on the market forces to supply them will obviously be in a weaker position. The Novelis and AcelorMittal companies recognise this. This is the trend for the future.

The sheer geographical scale of Africa is the challenge but the undoubted growth potential is the opportunity. Somewhere in the complicated maze of riches, violence, war and corruption its leaders must not be tempted by the quick fix and they must resist the blandishments of outsiders to take their toxic detritus as much as their high-tech but inappropriate solutions.

The continent of Africa has a special interest for me: that's where my roots are, where I was born, in Tanzania or Tanganyika as it was known at the time. Africa today is in some ways totally different from the Africa of the 1950s and 1960s – economically and politically – and yet for many it has barely changed as they face the toil of hardship and poverty; even in the rainbow nation of South Africa post-apartheid, thousands are still waiting in dilapidated townships for the promised transformation of their lives. There is increased prosperity and consumption in the major cities and that has brought with it all the challenges that the West has seen in the last half century.

The Western economies have experienced industrialisation but it has been a staged growth while the emerging nations are influenced by foreign direct investment (FDI) and have seen a rapid urban growth which is moving at a faster pace than the Western economies had to cope with all those years ago. Increasing urbanisation, increasing consumption, increasing demand for books, schools and hospitals all mean increasing waste, but the infrastructure and basic services are struggling to keep up or worse are being neglected. So while economies are booming thanks to FDI we must not lose sight of what we are doing with our waste and how we utilise it, whether it is for recycling, energy, or land filling. Whatever we do we must do it in the most scientific and economical manner.

Africa does not have a centralised approach for the continent and of course lacks the legislative disciplines imposed on Europe, so every country or small group of countries are tending to have their own drivers for recycling. However for recycling to work as we have seen in Europe some form of coordinated rules have to be applied without falling into the trap of over-regulation. Any form of centralised agreement is probably too far-fetched for the region but at least there might be a coordinated approach in terms of education and instituting collection systems which are user friendly and recycling friendly.

Recycling in under-developed countries is really not so much driven by the green agenda as much as by the economic agenda. The population particularly at the lower end will be looking to recycle things that can generate income: for example collecting metal, brass, or aluminium cans so they are able to sell them at a reasonable price to other middle men. There may not be a big enough incentive to collect waste paper because there are probably not enough centres to receive such material and if they exist they may be unable to process, sort and generate grades of recyclable material that can be used by the manufacturers.

But the market exists in Africa for the reuse of material within a country or intra-use within the continent; South Africa for example has a thriving paper industry and Nigeria has a good demand for paper because they have some fine paper mills. So Africa is the next challenge which many would like to be part of and support. There is room for intra-African trade and what they cannot use I am sure will find markets elsewhere.

The Middle East has so much wealth; Saudi Arabia can afford anything it wants today but it should use its wealth wisely recognising that it is the future it should be preparing for. Just because so much can be discarded and easily replaced does not mean that we should simply squander what we have.

Asia has been in one sense the lifeline which has helped the West recover from the worst recession in modern times and it too has an opportunity to build a solid recycling system from household and commercial streams. Over the many years I have been travelling to the region I have witnessed its phenomenal growth and it is clear to me that governments throughout the region recognise the situation. Their insistence on quality imports of recyclable material proves the point and has even provided the negligent West with a wake-up call. Standards must be raised and we do that by education, by the realisation that the developed nations cannot go on as we have been doing for the best part of 50 years.

13

Trading game

Imagine a juggler constantly trying to keep four or five balls in the air while others are throwing additional balls at him when least expected; then one begins to understand what global trading is all about in the recycling business.

When I started in the industry in the late 1970s, life was not so complicated. Back then the green phenomenon was not an issue and was certainly not having as strong an influence on collection from community streams. In fact the focus was largely on commercial collections. But very quickly, as volumes increased, it became a problem to find homes for the new waste. The volume of traded scrap metal was nothing like the level today and plastic recycling was virtually zero except for some of the finest quality, post-industrial material used for reprocessing. Paper for recycling then as now represented the largest amount of waste and the challenge even then was finding export markets other than domestic mills. The mills in Europe and America were certainly using different grades of fibre but not in the volumes we see today; nowadays some mills may be 100 per cent dependent on recycled paper whereas at that time the mix was likely to be 80 per cent pulp and 20 per cent recycling. So the pressure to find homes for the increased volume quickly became acute.

Europe and America started exporting to Asia in the late 1970s. It was a restricted market for paper in terms of grades and quality because the mills were not able to handle multiple grades. As a result higher grades such as computer printout paper, corrugated boxes and newsprint (i.e. post-industrial and not collected from the waste stream) were building up in volume and we had to find alternative markets which were willing and able to handle both the volume and grades. At the time export levels to India and China were less than 0.5 million tonnes compared with today's levels of more than 35 million tonnes and global levels of over 200 million tonnes.

The world has moved on and we are now talking of global trading because exports are important not only for the growth of countries but also to find a home for every kilo of waste that is collected in order to avoid landfill. Most Western countries are collecting more waste material than they can reuse although some grades of metal and paper being collected might still not be meeting domestic demand; that kind of variable will certainly continue to exist. But in principle most regions collect more than they can use so they need to export.

There are at least five major factors that influence global trading and each one has to be taken into account as their values rise and fall, sometimes on a daily basis (see Figure 7): logistics; currency volatility; global economic outlook; demand; and quality and legislative controls. Each factor is discussed further below.

Figure 7 Global trading

Logistics

This is the cost of moving the recyclable – raw material – from the point where it is processed to the final destination of the manufacturer which may be in Asia or anywhere else in the world. Export logistics can be split three ways: haulage at the origin, sea freight and inland haulage at the destination. Logistics constitutes a major part of the cost of recyclable raw material. Post-2008 we have experienced increased volatility in the freight rates impacting export costs (see Figure 8).

Currency factor

Another factor which greatly affects export costs is having to work with two or more currency variables. Post-2008 we have not really seen stability in the currency market for a sustained period; in the last five or six years to 2014 there has been a variation in the dollar against Sterling, the dollar against the euro, or with the Indian rupee, the Chinese renminbi, the Thai baht and the Indonesian rupiah. Because of the currency volatility costs have been in a constant state of flux and it is made more complex by the long delivery times. If an order is

Figure 8 Freight rates from Europe to Asia, 2008–2010.

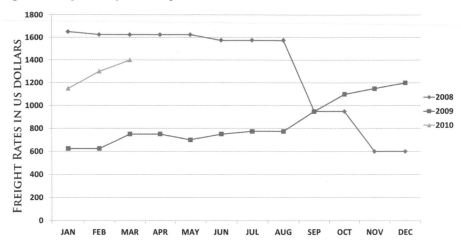

Figure 9 Chinese yuan renminbi to US dollar exchange rate, 1 January 2012 to 31 December 2013.

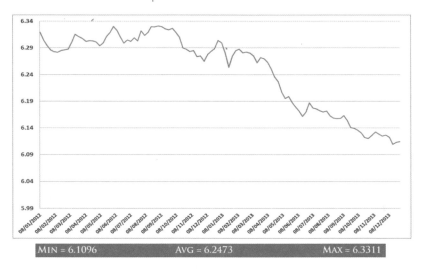

placed today by the time the cargo arrives many weeks later the currency may have moved 3–5 per cent up or down, which certainly affects costings. Figure 9 gives a snapshot to illustrate currency movement between the renminbi and the US dollar.

Economic outlook

The market is subject to economic moods swings as much as any other industry. As I write this we are emerging from a global economic slowdown which has been described as a double dip recession or even depression – some countries are managing the recovery process better than others. If the market suffers a slowdown it means products are not being sold which means there is less demand for steel, for boxes to package the new TVs or for plastic as part of the manufacturing process. While we may have a fixed cost for logistics, we can do nothing about currency fluctuations or the global economic outlook. Demand and supply are governed by the producer of recyclables being able to supply raw materials to the manufacturer who in turn is producing an end product for customers to meet their product demand. If more computers are being sold there will be a demand for more boxes to package them, equally if there is more construction there will be an increased demand for steel which means there will be a demand for scrap metal. The recycling industry is particularly sensitive to mood swings in the global economy.

Demand

One of the unique aspects of the recycling sector is its ability to predict in advance how the global economy is moving. The stocking levels provide a perfect barometer of storms on the horizon or of a brighter outlook as much as six months in advance. What is being collected and sorted today and produced as a recyclable raw material will eventually end up as a box for a new television, for example. But the time the TV is sold in a shop could well be six months away. So the box manufacturer, depending upon the signals received from further up the supply chain knows when to begin destocking or restocking according to perceived demand. The following could be a typical timespan following the movement of a recyclable raw material. The manufacturer allows for a 60–75 day journey and delivery time for the recycled raw material to arrive at the plant. This is followed by four to six weeks' processing time of the recyclables and thereafter conversion of the product into the final usable product such as boxes, beams or car engine parts. These products could be even further processed, for example the parts going into a car assembly. It is a very long term calculation but certainly a useful barometer for studying market trends.

When you peel off another black plastic bin liner for your waste it is worth considering just how long ago the manufacturer made his order for the plastic granules to be shipped half-way round the world before it reached your home as a bin liner.

So the raw materials at point zero that are being traded today will be converted into a consumable product in six months' time or even later and it was for that end point six months or more forward that the seller, trader, shipping lines and buyer all made their projected calculations. It is a long chain which makes the recycling industry a very advanced indicator of what the market is going to do in the future. Our supply chain tends to be longer than for those buying raw material based on 'just in time' delivery, which is more dominant today. Buyers are only ordering when they need products and are reluctant to hold large stocks. In addition while there are a number of major aluminium and steel stockists, there are far fewer of them compared with the number of recycling and collection streams around the world. A bearish or a bullish mood swing among these operators is a sure indication of how economies around the world are going to be behaving half a year or more hence. They are not just idly predicting for some analytical report, but making real financial decisions that are affecting them today. I would estimate that we are at least 18 to 20 weeks ahead of most suppliers or stockists.

Regulatory and quality controls

We have discussed the regulatory drivers and controls imposed on the industry. These regulations not only affect what and how we transport waste around the world but also have an impact when countries start imposing their quality controls which may mean qualities accepted in the past are now no longer allowed into a country. Just because they are forbidden does not mean that the particular waste should not be collected but what is required is careful handling and processing in terms of quality. Quality controls are likely to become ever more stringent.

There are any number of unexpected additional 'balls' with which we have to juggle and in early 2014 one of the trickiest was the weather. Severe storms swept across from America into Europe which disrupted shipping. Some vessels were already on the high seas, some were stuck in port but all the while the financial markets were moving and prices were shifting. Contracts had been signed, deals had been struck and no one seemed prepared to be flexible about this particular act of God.

How does all this uncertainty and volatility affect us as traders? We have got the hedging market where people are forecasting and hedging but as we are dealing in real time it is always a big challenge for us to make sure that we can find homes for the recyclable material that is collected across the various regions of the world. The challenge for us therefore is that at a very early stage we have to be able to create a model in our minds projecting where the market is moving. If we see there is a stocking drive because the global economy is anticipating an upturn according to people further along the supply chain we obviously try to buy and sell more. But there are times when destocking is appropriate, such as post-2008 and 2009, when market liquidity was very poor so although people wanted to buy tonnage there was no liquidity available to them. The consumer was worried about the financial meltdown so even consumer spending was down, industry players were destocking or carrying the bare minimum required to keep the business running.

The only upside in all these crises is that most industrial units can reduce production up to their breakeven point. Below that point it becomes non-economical for those plants to operate. If everyone goes down to the minimum level it certainly affects demand but for us as exporters the challenge is finding a home for the product. Then the home not the price becomes important. When in a stocking mode it is not the price that is most important but the availability of the product and being able to find space on the ships so that the cargo can be exported in good time. That becomes the real driver because during such times the shipping lines are also trying to sell their space at a premium. But we can only absorb freight and haulage costs up to a certain point beyond which it becomes too expensive for the reprocessor to buy that material.

We are constantly juggling all these balls in the air: are we in an upturn or a downturn? Where is the currency going to be? What is the projected demand? How will the logistics costs behave? Are the shipping lines busy? Are the recyclates going to be in demand because we are always trying to price ourselves in a competitive market? We must constantly assess all these factors and set a price that works for us as exporters, for the producer and for the buyer ultimately so he or she can reprocess the material economically into an end product to sell into the market.

Equally a producer of waste is on a contract to receive agreed volumes of waste from commercial or community streams into the processing plant 24 hours a day, every day. The operator often takes this material at a fixed price which may have been fixed six months or a year earlier when the market may have been badly judged or the operator may have been forced to buy the material at the highest tender price to secure the waste to feed the plant.

The industrial demand for recycling material continues even in a downturn because the mills or the reprocessor will always be requiring raw material to run their plants. Depending at what end of the process one is the recycling plant will always offer tonnage for sale on a weekly or monthly basis to the producer. We traders in turn will be sourcing recyclables for export around the world.

In the last few years the competition for both buying and using this material has increased tremendously. We certainly have seen many more export houses emerging in the last five years to 2013/14 who want to have a stake in the recycling business because modern plants are more geared to using greater volumes of recycling material than virgin material therefore the demand has increased making the industry more attractive.

There are different levels of traders in the market from the very small SMEs who might be shipping a few containers a week to a very large global trader who could be moving whole shiploads of metal scrap across the world. The range in the types and sizes of business houses that are trading in recycling material varies across the spectrum. Most people tend to be in the middle of the spectrum trading about 100,000 tonnes of recyclables per year, then there are the very large businesses, supplying over 250,000, and the smaller ones, less than 50,000. Most of the global traders would be between 50,000 and 250,000 tonnes.

The challenge is the same for all these houses; that is the ability to source the tonnage from recycling plants at the right price, at the most competitive logistic cost with currency exposure fully covered and having a payment instrument for the material shipped so that payment is guaranteed.

For the payment instrument, different grades trade differently. Some people operate against letters of credit which can be of different types – payable at source, payable at destination, payable on sight or at sight, payable at the end of a certain period be it 90 days, 120 or 180 days – known as USANCE Letters of Credit. Then there are people who will pay 20–30 per cent deposit against the order with the balance to be paid when goods are received or even after they have been delivered and checked at the destination. Goods can be sent on credit. Businesses may even put goods on the high seas without an order, trying to find a buyer while the ship is en route. In other words trading is not just a case of moving cargo but really a whole raft of challenges.

Unfortunately exporters the world over in the past did not consider the buyer's own issues in terms of the buyer's country's requirements that seriously. We always felt that buyers should accept what we were selling to them. But over the years the buyers have been saying: I don't mind accepting what you are exporting but you must also have an understanding of what is allowed in my country and what is not allowed. As a result we have seen the Green Fence controls in China which helped the industry tremendously because it brought home the message to all of us that exporting to China did not only mean selling to them what we were producing but selling them what we were producing so long as it met the requirements of the government of China. In other words what crosses through the borders of China or elsewhere for that matter must only be waste that is allowable by those countries. The problem has been that we at our sorting plants have never educated our pickers and sorters about what should and should not be included in the material for export to China. Similarly material for India must also meet the requirements of the Government of India. The same goes for Korea, Indonesia and will apply more widely to parts of Africa. Accordingly only the material that meets the requirement should be shipped and if it doesn't it makes no sense to buy or produce it because there will probably not be a home for the material.

As emerging and 'emerged' markets open up there will be new opportunities as well as a further set of regulations for the recycling industry to consider. The Brazilian economy is growing and consumption increasing. I can see Brazil growing steadily to become an export partner. It will be producing more material than its own domestic market can consume so it will be looking for export markets. On the other hand when I look at South Africa and India their demand for recyclables is increasing tremendously because they need more material. In the case of South Africa I know that it does produce certain grades of recyclables in excess of what it can consume so it has some exports to neighbouring African countries or even as far as India. But India is totally dependent on importing its recyclables because its own collection streams and recycling, whether in steel, plastic, non-ferrous or paper, within the country are far from meeting the demand; therefore it will remain a net importer.

As for Eastern Europe, in the last few years I have seen exports out of Russia going to Asian markets. I have seen Poland, the Czech Republic, Hungary and Croatia all starting to put in place collection streams for both commercial and community waste in different grades of recyclables which means they are also coming on the market as suppliers of recyclables. But they are also coming on the market as big consumers of recyclables as new industries are springing up. Poland for example has been undergoing rapid industrial growth with many new industries coming on stream and capacity has increased. I know specifically of a new paper plant which has come on stream to meet the growing Polish demand and will therefore consume most of Poland's paper for recycling whereas in the past this surplus material would have gone to Germany.

All these economies are introducing new plants or new collection streams which are producing more recycling material which either means they have more material for export or less material requirement for import because they are generating more at home. This adds to the challenge of competing globally for a home for the recyclables, which in turn makes global trading more interesting and demanding.

Global trading in respect of recyclables is a unique challenge because trading in any other product is much less tightly regulated. The challenges of moving one tonne of plastic scrap, or steel scrap compared with moving a tonne of shampoo are quite different and much more complicated although both are global trading. Trading in recyclables should not be misconstrued as trading in Coca-Cola or cigarettes or shampoo from one part of the world to another. There are many more factors and one of the trickiest is the question of trust.

With the volume of waste that travels around the world in containers it makes it practically impossible to have every single box inspected other than when it is checked on arrival. Therefore in this global trade there is a lot of trust and understanding down the chain but what is lacking is education. There may always be trust but if there is no understanding of what China will or will not allow trust is not enough. The reprocessor may have done his or her best but this may not be good enough if there is a lack of understanding of China's requirements. So the question is always: what's in the box?

14

The container journey

Trading patterns in any business are always challenging in that one has to have material to sell, one has to find a buyer and then one has to have the necessary infrastructure and logistic arrangements in place to pick up the material and deliver it to the end user. In the waste recycling game there are many other constraints. In addition one has to be sure to meet the necessary regulatory controls and the licences required to transport the material by road; one also has to meet other regulatory requirements such as forms to be completed to travel with the cargo which meet the environmental regulations of the country in which one is operating. In the middle of all that there are also inspection regimes from the buying countries. Will the product meet the quality standards required by the recipient country? And finally will you get paid?

Let us go back to the beginning to the point where the householder sorted the waste at home and put it into bags for collection. It is collected and brought to a transfer station or recycling plant where it is then sorted again into various grades and baled. Now the product is ready for sale which makes the process sound straightforward enough.

The waste being sold comes in multiple shapes and sizes. It could be briquettes of metal, loose scrap metal, baled paper or plastic. The briquettes of copper may be very small or scrap metal could be bulky. It might be cut into lengths such as rails or building RSJs (reinforced steel joists) as required by the final customer. Some of the waste may be shredded material, which could be steel, tyres and textiles; it might be baled like paper and plastic or glass which is loose.

All these recycling materials in different forms are made ready for transportation for their journey to domestic destinations or export to international markets. If the product is moving domestically, it is normally sent by road or very rarely by barge or train. Sales within Europe from the UK are normally sent by trailer or even by container particularly if they are heading to destinations as far as Turkey. Sales outside Europe normally go by large shipping vessels and the most cost-effective method of transportation is containerisation. But long before the waste reaches its final destination there are multiple hurdles to be overcome all of which impact on the quality, price and ultimately the many recycled products which end up back in the shops.

I shall focus on container shipment which is the bulk of the daily trade. Firstly, special procedures demanded by the recipient country have to be followed. Secondly there is local legislation governing waste movement in the source country before it reaches the port for loading. Legislation is also very specific when waste is moving outside the country – we call it TFS regulation (Trans Frontier Shipment). Anyone who moves waste within the European Union must have a waste carrier licence and people trading in waste around Europe need to have a waste broker's licence issued and authorised by the UK Environment Agency.

Thirdly, the waste that is being moved must follow certain quality constraints and certain rulings from the Environment Agency before it is loaded at the point of departure. Not only is the container liable to be checked at the facility where it is being filled, it is also liable to be checked at the point of loading onto the ship at the dockside or last departure point out of the country. Environment inspectors and enforcement officers from the country of destination regularly check the waste that is moving out of the country to make sure that it meets the international guidelines regarding the movement of waste and meets their own domestic requirements. Certain countries require inspection for quality control to be done by their own authorised agents in the exporting country. For example China has introduced an inspection called CCIC, China Certification and Inspection Company. The company was founded in 1991 and is an independent, third-party certification operation. Without their certification the freight cannot be shipped. All of which is an attempt by China to try to control the waste that flows into the country, to ensure that it meets the Chinese Government's requirements, is a product that has a use and is in demand in China. They want to be certain that the shipment does not contain any radioactive, toxic or non-usable waste which is only being shipped to China for burning or throwing into landfill.

Therefore not only does one have to meet a range of domestic standards all of which can be verified at any point but one has to be certain that there is nothing lurking in the container which is prohibited by the recipient country. This is where trust comes in because it is impossible to verify the contents of every container one is shipping. This of course means one has to have a detailed and current knowledge of the requirements of the recipient country which increasingly have been strengthened and enforced in the last few years to 2014.

China is not alone in seeking to protect itself from illegal shipments. In India there is also a pre-shipment inspection control of the waste that goes into the country by inspectors or agencies approved by the Indian Government. In the case of metal it must also be certified free of radioactivity. The exporter carries the responsibility that the quality of waste that is going into India is within allowable standards imposed by the Government of India.

In the same way Indonesia is gradually bringing in its own controls which state that they need all cargoes coming into Indonesia to be pre-inspected by government appointed agents who inspect the cargo to make sure it is of the right quality, the right specification and meets the standards of the government.

These are all broad standards which have really helped growth in the industry of good quality material being processed for export not just processed for the sake of processing. I am in favour of controls and I am very much in favour of Green Fence standards as imposed by China because all this has driven home the message that increased volume of recyclables is important, but more important than volume is quality. I have always opposed excessive legislation as it starts to hinder the movement of trade; but reasonable controls that bring home the quality message are good.

Of course all these checks require a certificate proving that agencies have inspected the material and that it is worthy of the countries to which the cargo is destined. Not only do we have to arrange the inspections but it is vital that the certificates are included in the international documentation that accompanies each container.

Assuming therefore that the truckers and the exporters and the transfer stations have their various licences, the inspections have been done by the right agencies, the quality has been approved, the material is authorised to be loaded into the containers and the normal customs declarations have been made, the waste is ready to start its journey for export. Thereafter the container is put on tractor units which then carry the containers from the loading site to the port or to the nearest terminal. Containers tend to arrive at the port within a prescribed number of days before the vessel actually docks because there is a huge demand for space alongside the berth. Until such a time the containers are stacked some distance away.

Once all the paperwork is in place the containers are loaded and the ship continues on its journey to its final destination. When it has reached its destination, the materials are discharged at the quay where the buyer of the material is also obliged to make sure the material that has arrived is correct. At destination imports are liable to random controls to make sure the material inside the box matches the import documentation which the buyer/importer declares at the point of entry to the country.

Assuming everything is in order the cargo begins the last leg of the journey to the processing centre, plastics or paper mill. At that point the material is discharged from the containers ready for use.

As we have mentioned the journey time from loading point to destination may take anything up to 60 days or more. That is roughly a two month time span from the moment the material is loaded in the box until it arrives at the other end.

When the box is finally opened it can be unpleasant. If the material that went inside the container was contaminated, dirty, contained food particles, plastics, Coca-Cola bottles with liquid still inside, milk bottles with milk inside, paper with sandwiches or chips thrown inside it will all ferment on the journey – another reason for us to ensure that what goes inside the box is as clean as possible.

Remember there could be another 15–20 days added to the journey time above from when the individual threw the waste into the bin at home and waited a week for the next collection plus another seven days for it to arrive at the processing centre or transfer station for its initial sorting and baling. That might be anything from 10 to 12 days before that milk carton or meat packaging gets into a bale form. So on average it could be as much as 75 days before anyone can start the actual recycling process of those cartons.

Security in the recycling sector is just as much of an issue as in any other industry. Over the years people have lost the contents of containers that might have contained more expensive material – metal briquettes, steel, copper. To ensure that the contents of the box remain secure each one has an identifiable seal with a number which is given to the buyer so the delivery can be checked when it is discharged. This is obviously critical with high value items. Also included in the documentation which has to be presented at the port of loading is a set of photographs showing the different stages of the loading operation: when the container is empty, when it is half loaded, when it is full and when the door is closed and sealed.

Because inspectors cannot physically check every single container they rely on this photographic evidence. Some inspectors choose to check 5 per cent at random, some 10 per cent, but the inspectors decide which box, when and where. They can judge from the photographs what has gone in the box and what the exporter says has gone in the box based on the documents presented. If the exporter claims to have loaded copper and the photographs show steel scrap that is an obvious discrepancy, sometimes it can be more subtle such as in different grades of paper.

There is something of an art in loading a container. Obviously the aim is to fill it to the brim to occupy all the space in the box and maximise the payload. If the goods do not fit properly into the box then naturally you are paying for less weight and more air. The shipping line collecting the freight imposes a maximum payload per container and not a minimum. So the lower the payload the higher the cost per tonne.

Increasingly now balers are producing baling equipment which ensures that the maximum weight goes inside the box whereas 30 years ago when we used different sized bales achieving a high payload per container was always a challenge. In order to achieve the maximum weight specialised equipment is used to compress the bale because the total weight that goes inside the box will be weight per bale multiplied by the number of bales. Therefore increasing the number of bales and the weight inside the box will decrease the freight cost per tonne and increase revenue per box.

But even today there is variation in bale sizes and weights where this concept is not understood so it becomes more important that investment is made to ensure that people who move waste round the world do it in the most economical way. Not only does it increase profits but it also reduces the carbon footprint. We want to move the maximum weight allowable with minimum energy expended.

When I started in the business the concept of container loading, namely receiving, loading and packing the containers was not the norm. The usual mode of transport was putting waste in the back of a truck. To load a container three basic systems are needed: a ramp to go from ground level into the container; a fork lift with a boom length able to carry the load – for example the fork lift is designed to be able to carry one bale or a few briquettes of metal right inside the container or if they are on pallets to lift them into the box; and the ability to ensure maximum loading into the box. In the past people didn't have ramps or enough volume of different grades required for export so my first exports were made possible by creating consolidating centres – I established one of the first in the UK in Purfleet. I had suppliers who would bring their waste in 3, 5 or 6-tonne trucks. We would receive them, consolidate the load there and the warehouse would have the necessary forklifts and ramps to load the material onto the container. Today as volumes collected have increased it has become more economical for all these plants to have their own ramps and forklifts so they could do the loading themselves. Now a container can be packed in less than 20 minutes because the bales are the right size, the fork lift comes in and 40–48 bales are loaded into the box which is then ready for its journey.

As we have noted elsewhere amid all this activity there are illegal exports. In the past there were no regulatory controls in Europe or indeed imposed by recipient countries – there are still very few controls imposed by many African countries although Asia has moved on quite a long way in last 20 years. People who collected and sorted waste probably felt any kind of waste must have a home and once it was loaded in the box and going to destinations where there was cheap labour and large industries, someone somewhere or another would have a need for it. Today exports to Asia need to meet the quality standards that are no different from what is required in America or Europe, the raw material in any part of the word has to be of the same quality. The grades may vary but everybody needs the recyclables to be of similar specification.

Over the years it has become necessary to impose controls to ensure that waste moved around the world was a raw material and not rubbish for dumping. Controls today mean it is very difficult to move material that does not fit the bill, and also legitimate exporters are not willing to risk shipping material which does not meet specifications.

As the world has grown to be carbon aware the shipping industry has moved in the same direction which has also impacted on trade. The larger ships used to run at a faster speed but now they have slowed down because they have calculated that if they reduce their speed it is more fuel efficient. Most ships when they berth now run on fuel efficient supplementary engines and are not powered by the ships' main engines when they are at the quayside. The carbon footprint per metric tonne of recyclables moving on the high seas today will be much lower than what it was at the start of the 21st century. Slower journeys however inevitably increase the time it takes to get our waste shipments to the buyer.

There is an additional obstacle relating to the letter of credit we discussed in the last chapter. These letters are not only clearly linked to the cargo but they are also time specific. The letter states precisely the date by which the buyer needs the goods to be shipped. As we have noted not everything runs like clockwork.

As traders we have to notify the inspection authorities the loading plan showing the loading sites and dates of loading. This has to be done three or four days before confirmation with the shipping line because they also need a cooling off period of 72 hours. The inspector will look at the schedule, see what else is going to his or her country from the exporting country, and then chooses to inspect certain boxes. According to whatever criteria have been adopted, the exporter has to ensure the inspection is carried out as required. We then have a duty to ensure that our containers do arrive on time for the inspection to be carried out. If we are late the inspector will be gone and then there is a cost because our containers may miss their ship's departure time and consequently the Letter of Credit may be invalid.

Similarly the weather can play tricks. A certain cargo might be booked on a certain ship, but because of the weather the ship may have to set off early and normal load time is reduced resulting in either some of the containers being loaded or none. This could cause a problem in meeting letter of credit stipulations and therefore may result in non-payment due to discrepancies. These are some of the challenges which are making global trade increasingly difficult which is why over-legislation does not help the industry, rather it hinders growth. In my opinion before legislators add any new rules on any point of law they must see what legislation already exists on the subject. By taking that approach they will help to sustain the business.

Generally everything goes smoothly as long as a structured model can be followed which takes into account every part of the process. But taking short cuts or trying to do it faster and better will always lead to problems and there are people who do end up having great difficulties. We always stick to our own tried and tested model where we take into account every stage of the export chain management and ensure that every step is correctly followed, respected and satisfied trying to anticipate every eventuality. This requires current knowledge of recipient countries which is best achieved by regular face-to-face contact with the buyers.

Occasionally *force majeure* conditions such as the bad weather in 2013/14 delay containers from reaching the terminal on time or an accident or vehicle breakdown or a train carrying the cargo may be delayed. Bad weather can delay ships' sailing times or the ship herself may have engine problems in port. Shipping lines don't like running behind schedule so to make up for lost sailing time it is perfectly common for the vessel to reduce loading times or cancel docking at a particular port bearing in mind that a ship has to dock at many ports on its rotation. The only good thing now with these huge vessels of 14,000, 16,000 or even18,000 teu (twenty foot equivalent units) is they have reduced the number of port calls they make in Europe as their cargo is distributed using smaller vessels or rail or road. But if the vessel has been delayed on the way perhaps passing through the Suez Canal or has a breakdown then the vessel planners thinking ahead have got to cut out one or more ports to make up for lost time. In other words if a vessel decides to omit a port in the UK from which the cargo was designed to leave at the end of the month in order to meet the letter of credit expiry dates, then the shipping line may put the cargo on the next available vessel. But that may now have defeated the letter of credit because the date of the last day of shipment has passed. The shipping lines always say they will endeavour to put a cargo on the ship but they offer no guarantees.

All this information must be weighed up when working out a loading plan with dates all along the line. If suddenly dates are not being met an alternative must be found, negotiating an extension on the letter of credit or switching cargoes to a shipment which has a longer deadline.

These are the trading scenarios that must be faced every day and which have to be linked all the way back to the offer made for that first bale of waste.

15

Zero waste and future challenges

The problem is we can't say no. We always want more; we have to have the latest gadget, the latest fashion, the newest car, even another helping. Since the beginning of time we have been acquisitive, the only difference is that once we needed things – a bigger club, a warm animal skin – to survive, today we simply want more precisely because it is new or the latest fad. In some parts of the world of course acquiring things is still a question of survival but once that threshold has been crossed it is not long before the need to have more kicks in; mobile phones are a must have in parts of Africa where not so long ago one daily meal would have been the only priority. The boom times which have swept through the major cities of India, have delivered untold wealth to a tiny percentage of the population while the vast majority remain in poverty watching in awe, envy or despair as fleets of expensive limousines sweep past. This is not a moral point about the haves and have-nots, it is simply stating the obvious that as soon as we are able to we will always want to acquire more and therefore as more people, particularly in the developing parts of the world, move from poverty to modest middle class and higher they too will consume more and begin to discard more. There is nothing that can stop this relentless progress. When once we repaired what was broken, we now simply replace it with a new model. Waste is the only certain outcome of this growing prosperity.

The aim now of the affluent Western world is to stem this tide and achieve zero waste which means trying to stop any waste going to landfill. It was a term coined in 1970 and the idea of zero waste runs in parallel with the philosophy of the 3 Rs – reduce, reuse and recycle. Some say there should be another 'R' before those three: Refuse. By turning away from the inclination to acquire more we might end up by discarding less.

But it is the job of the recycling industry to deal with the reality of life, not to preach, and for us that means collecting, sorting, processing and devising cleverer ways to handle the mountains of waste that we all throw away every day.

As the concept of the three 'Rs' started gaining ground, zero waste perfectly matched the dream – if we reduce our waste and reuse or recycle as much as we can then we should come close to doing away with the need for any waste going into landfill. Gradually over the years the idea of burning the residual waste has gained ground, technology has improved, waste-to-energy techniques have been devised and RDF has become increasingly popular, particularly as energy costs have risen so dramatically. So zero waste perhaps has become more than a dream – it might even be a tangible goal – but only in developed countries.

Western economies believe fervently in zero waste but in countries where basic recycling has not made enough progress or maybe even started, or where targets have not been set let alone achieved, zero waste is a long way off. One could say a short cut is to burn everything, but then the raw material is destroyed forever; whether fibre or plastic or any other waste, these are precious raw materials. The problem of course is that while the use of mobile phones and computers is multiplying rapidly in emerging countries, recycling there is not keeping pace, the facilities largely don't exist to extract the precious metals from the motherboards or even simply the plastic casing.

Some things cannot be burned such as electronic waste and more worryingly toxic waste. Chemical waste and medical waste must all be carefully treated before disposal to reduce the toxicity and so alternative methods of disposal have to be found.

So zero waste will depend on the type of product, the type of economy and the types of collection, processing and possible uses one has at one's disposal. Alternatively one can ship the material off to another country which is able to process it. All these factors will govern whether zero waste can really be achieved globally.

Outside agencies also play an important role in helping the recycling industry achieve something approaching close proximity to the zero waste target. In this context many manufacturers are demanding that packaging suppliers offer packaging which is recycling friendly and easy to dispose of and collect. Retailers want to reduce the volume of packaging required in the course of manufacturing and increasingly we are seeing many products which are biodegradable. This is an attempt to remove what one might describe as the design flaw in manufacturing. By building in a natural degradation we can cut the amount of waste that has to be treated.

On the other hand in our pursuit of clever design we inadvertently create additional problems. One very simple example is self-sealing envelopes. Some of the new glues are becoming more difficult to process. Glue is normally water soluble, but in cost saving drives, environmentally friendly initiatives and a desire to reduce the volume of packaging the new glues now being used, which have better adhesive properties and are therefore better for packaging, are harder to remove from the paper which in turn adds to the cost for the recycler to process.

Technology therefore is constantly changing and producing new products and as a result presenting new hurdles for recycling. The challenge for achieving zero waste is to increase the volume being recycled, whereas the challenge for the manufacturer is to produce the best possible product which also increases recyclability, lowers cost and creates less packaging. The two are running in parallel to reach zero waste but reaching that goal will become harder.

The question is: are the waste targets we are setting ourselves simply too difficult to achieve? As we have touched on earlier moving from 50 per cent recycling is achievable but it becomes much harder moving from the higher levels of 75 per cent and beyond. The cost of increasing the rate by just 1 per cent does not always make economic sense. At that point it is the quality of the residual waste that is important and is the extra recycling effort required justified or should the excess simply be burned or directed to landfill?

Do we have to accept that the future will always mean that we have to make use of landfill sites and we should just accept that reality? As the global population increases and we are all consuming more, as medical advancements for example take place and chemical industries come up with new products, we in the recycling industry are also faced with new material that needs to be recycled. Just as we will always be creating waste, not all waste can be user friendly in terms of recyclability. We will always have to find ways of treating waste before landfilling for example detoxifying chemical waste, reducing the volume of waste by burning it, shredding it or granulating it so that the volume of space required in landfill is reduced. We will need to find a home for these products so we will always need landfills. What is important is that modern landfills are more environmentally friendly, they are dry sealed with multi layers of protection so the permeability of waste into the surrounding soil is reduced to as close to zero as possible. It is a balance between environmental protection and economics, assessing the benefits and the risks; what savings can be made by burning the residual waste as an alternative source of energy and reducing the use of fossil fuels? How much is lost by burning waste which could be recycled? We will get closer to zero waste but we can never achieve it.

So what are the future challenges for the industry as it battles to attain seemingly impossible dreams? Francis Veys, former Director General of BIR, summarised how he sees the industry developing:

> Because of the global need for raw materials, for energy and for environmental protection the Recycling Industry has now become and will continue to be a key sector in the world economy.

The sustainability of the product which is put on the market and of the human behaviour will continue to be the leitmotiv of the 21st century. However as the world economy is directly dependent on geopolitics the development of recycling will mirror the macropolitical changes in our planet.

While we will reach the limit of growth in recycling in the (old) industrialized countries, the recycling sector will continue to develop itself in re-emerging economies – China, India were key players in the world economy more than a century ago and they are re-emerging rather than emerging – and literally 'start and boom' in new emerging countries. While the 'old' western world will essentially need to specialize itself in 'niche' markets, in sophisticated materials and hi-tech sectors and hence reduce the production of 'simple' goods which will be better manufactured elsewhere at a lower cost, the (re)emerging markets will use 'basic' primary raw materials and recyclables to manufacture less complex products to develop their new infrastructures (bridges, buildings, roads, machinery) and offer to their citizens the questionable 'joy' of consumerism (cars, household appliances, A-V's, magazines, packagings).

This trend which already influences the international flows of raw materials – primary and secondary – but also of energy for almost a decade should accelerate until 2020 and beyond, unless some major catastrophes – natural (earthquake, flood) or artificial (wars) modify the geopolitics map of our today's civilisation.

Obviously the difference between the current 'old' and 'new' worlds will progressively fade away as soon as the new emerging markets become industrialized and start specializing themselves to manufacture less basic products. In addition, the globalization of the companies and of the recycling sector will increase the number of multinationals which would have a direct impact on the development of the industry and of the market worldwide.

Re-actions:

we will see growing market protectionisms from countries looking at securing their primary raw materials/recyclables and energy (trade barriers, environmental legislation, technical barriers, certifications up to a nationalization of the sector).

Pro-actions:

we will see closer cooperation between the producers and the recyclers, growing investments in R&D to improve the duration, the quality and the recyclability of the products/goods, to avoid impurities, to avoid potential 'waste' at any stage of the production and consumption, and to increase the volume of recyclables.[81]

---

81 Francis Veys, personal communication

The investment Francis Veys speaks of will depend on the harsh reality of what returns can be achieved. Prices of recyclables have been fluctuating over the last 20 years. In the past the available volumes of recycling material were lower. We have talked about China and its phenomenal growth. Western economies' consumption has also increased but the increase is nothing compared with China which in 20 years has reached the same level of demand that took the West 60 years to attain.

During this period of rapid growth in demand for recyclables the price fluctuation has been very strong. We have seen many peaks and troughs in last ten year cycle to 2014 in particular. Across the board the prices have risen – although the global recession and financial meltdown in 2008 has had a cyclical effect on finished goods which has had an impact on the price of recycled material.

However, there is certainly a continuing upward trend in prices but we will probably not see some of the dizzy heights reached pre-2008; scrap metal was a luxury only enjoyed by the Western economies, today every country including Africa is collecting scrap metal. So will we again see the price of steel scrap shooting up to levels of $800 per tonne and higher? Will we see the price of paper for recycling going to $350, or the price of pulp to $1,100–1,200? Copper prices touched $8,000 pmt, platinum hit $54,000 per kilo. Will these levels be breached? What will happen to rare earth metals? The volume of rare earth metals collected for the recycling industry is definitely not of the highest quality or quantity but will we also see a price explosion with these metals? How far and by how much will these prices fluctuate? In the future are we going to see a growing supply being closer to the curve in demand we are anticipating in the next 25–30 years or will supply still be lagging behind demand?

Personally I don't see volatility subsiding and we will probably see similar trends in future. The question is: will the volatility see us breaching the levels we have seen in the past?

The concept of recycling which was restricted to the Western economies is now a global phenomenon. We have seen the recycling industry grow worldwide. The volume of recyclable material available is going to grow in line with increasing demand. The demand will increase as we have noted earlier as a result of growing population and by growing consumerism. So these drivers will increase the pressures but at the same time the recycling industries which in the past were restricted to the US and Europe have expanded to Japan which became an exporter post-1995. Australia is now exporting as is the Middle East and China has also been increasing its domestic collection (over 50 million tonnes in paper). As the demand increases so too does the availability of recyclables.

Is it scaremongering to say we have finite natural resources and ultimately they will run out? We all have read that the speed at which the global population is growing and emerging economies are increasing their consumption of clothing, medicines, cars, mobiles and food means we will simply run out of space, never mind natural resources. I am not so pessimistic but it is a timely warning. Mother Earth has provided us with enough raw materials for basic living but we are exceeding that modest level. We have multiple cars in one family, we enjoy the luxury of flying at cheap rates at the drop of a hat and we insist on goods regardless of the season which have to be shipped round the world. Taken together some might regard this as profligacy on a gargantuan scale and probably the resources at our disposal are not going to meet demand.

But the human mind has come up with the brilliant idea of recycling, technology never stops advancing, we are all in some way contributing to and promoting recycling whenever we dispose of our waste in bottle banks or sort our plastic bottles from our glass at home. On a commercial and industrial level the recycling process is built in to every business plan. So we are supplementing our supply of finite resources with recyclables; therefore we are not using as much of the natural raw material which will allow us to expand our lives. But we must take care not to keep demanding more as I said at the start of this chapter. Refusing to say 'yes' in the first place is more important than having to recycle at a later stage.

From an industry point of view I am certain the future is bright for the very reason that we will continue to have to devise new ways of disposing of waste. The industry must continue to develop new methods of reusing recyclables which reduce the stress on finite resources. Our real challenge is how we can harness the qualities of the manufactured products in order to reuse them at a more competitive price by using less energy, having a better carbon footprint and at the same time always being able to make our final products biodegradable or recyclable.

In the end it will always depend on economics – the dollar will probably always beat the environment. The driver for the industry is satisfying the legislative requirements and zero waste ambitions aimed at reducing our carbon footprint. All this becomes the primary driver but what drives development and innovation, what puts industries and businesses in the race is the dollar. The bottom line will always be the motivator perhaps inspired by environmental concerns but ultimately what we do has to make economic sense in order to create and expand an industry which is sound and robust. It is high time for some radical thinking.

16

New approach

There is probably no other industry in the world which touches every life every day. From pauper to prince we are all in some way interacting with the waste recycling industry. We might be choosing a new sports car or we might be scavenging over a rubbish dump for the wherewithal to survive the day, but in some way to a greater or lesser extent we are all distant, global partners in this enterprise. We can't do without it but we want nothing to do with it. It is a business many would prefer remained on the other side of the tracks. It is high time waste was recognised for what it is – a highly valuable raw material, even a precious raw material, which we depend on to get through our day whether in air-conditioned comfort or abject poverty.

In this book I have talked about increasing collection and increasing quality demands. Both of these are becoming more and more important because the more we recycle the less ends up going into landfill. By producing a usable quality raw material we are able to promote its usage by manufacturers, reduce costs and of course preserve finite natural resources. We have also been talking about increasing legislation. At its best laws help and protect the industry; help us to generate precisely the quality we should be striving for and protect us from illegal activity which only succeeds in making the global movement of waste harder as restrictive laws are introduced to combat the practice.

But for us to succeed in increasing recycling and improving the quality of the material we trade in, requires a major shift in thinking on the part of all governments and that is an agreement to act in unison, to have consistent and coherent rules which can and should be applied the world over.

While acting with the best intentions, all kinds of waste unfortunately are classified differently in every exporting region of the world; everyone creates their own standard which actually defeats the object of the exercise. It destroys the basic understanding of quality meeting a specification because no two are alike. The Americans have their own specification, the Europeans have their own standards, even within Europe we find individual countries have their own local rules apparently aimed only at matching the qualities in their own country with little regard for even European-wide specifications. The same story can be found in India and China.

While the intention to be tough on quality may be laudable, this diversity is not really taking the drive to increase high standards in recycling forward. I don't underestimate the challenge to bring about uniformity; it will take quite a few years to get agreement as there are so many different issues at stake. Nevertheless we must have a globally accepted quality specification for all grades of waste that are produced in all recycling sectors if we are to achieve what is essential – a global recycling loop. This will help the producer of the waste and the reprocessor of waste to reach the quality that is required.

In fact there is not such a wide difference between us; sometimes it is only a slight variation in the specification but that can be enough for a waste product to be rejected because the description is not recognised or may have a different name even though intrinsically it may be the same product.

This method of describing what we produce is therefore fundamental for the industry and I would say it has to begin with the USA and Europe as the two major producers of recycling material across the world. At the very least these two regions must create joint specifications which are enforceable in both parts of the world. Everyone is now demanding quality and high standards but if we cannot agree on what constitutes a particular grade of paper what hope will there be in finding agreement to define hazardous waste.

This brings us to the second essential of global legislation. It is vital for governments to work together in helping to promote safe passage of internationally recognised recyclable grades. Waste may be classified as hazardous because it does not meet with specific export requirements of a country: for example when we send metal scrap to countries such as India we need to provide certificates stating that they are free of radioactivity. But the movement of the same metal scrap may not require similar documentation within Europe. However at some ports scrap metal passes through electronically controlled entry points where the shipment is tested for radioactivity.

The point is we need consistency and testing for radioactivity is just one example. We must have a coordinated approach so that safe passage of waste is encouraged to help promote trading of waste in our global loop. What is important to remember is that a tonne of waste may be surplus to requirements in one country but it could well be in very short supply elsewhere; therefore this waste must be able to move freely for the good of the planet. All customs authorities across the world should recognise the same standards and procedures and provide the same degree of controls as each other.

I can understand how some of the discrepancies have arisen. Some countries have found that they have become the dumping grounds for other people's waste. Suddenly aware of the dangers, they have introduced tough measures and may find that they have over-regulated. Other countries may have a more relaxed approach and be under-regulated, content to take in perhaps dubious even risky cargoes, while in the middle you find the happy medium. However the consequence of this differential makes the free movement of waste around the world more challenging. Uniformity in regulation should be our endeavour; it will be difficult but we must try because we have a duty of care.

Waste that is collected goes to a plant but the facility may not have a state-of-the-art modern sorting system or may simply be outdated. The responsibility for dealing in waste that does not conform to internationally agreed specifications needs to be defined and policed. While I advocate consistency in legislation I do not do so to make it easier to ship unwanted waste. The movement of waste should only be allowed if it conforms to globally approved specifications but local authorities and councils must be held responsible if they fail to stop non-conforming waste getting out into the market.

People are all too ready to cut corners and it is still happening in 2014 because they think they can get away with it. We have seen in the past contaminated waste which is of no particular value, being shipped overseas instead of going to landfill. To me the exporter is responsible for moving what amounts to quite literally rubbish, knowing it to be worthless, just to dump on another nation's doorstep. The recycling plant selling the waste to the trader is shirking its duty of care; the trader that takes on that duty of care may try to ship a cargo knowing that it does not meet any recognised specification. Where does the responsibility lie and how can it be enforced internationally? If a recycling plant is producing sub-standard recyclate which is non-conforming then who should be responsible for controlling such production? The city authority for allowing the processor to continue in business or the processor, or even both?

The ultimate aim is that authorities giving licences to collect and recycle should understand the composition of waste that is being generated and even have some understanding of the requirements of the ultimate buyer – if they don't know they should seek advice. The legal maxim that ignorance is no defence should surely apply. They must find ways to ensure that the waste is only processed in plants that have the capacity to recycle to a better degree and produce more of the right quality of material that is required.

This is not a challenge to the smaller operators who have a key if differing role; history tells us what happens when monopoly positions occur in any industry or worse still authorities wash their hands of the problem.

Usually the solution to any problem and thus a new way forward can be found in an anomaly with an existing situation and that applies well in our industry. At last we are realising that one of our biggest mistakes was that we never fully understood the needs of the end users of our product and what they intended to use our recyclables for in their industry. The golden target, the great rush in the industry, has always been, where can I get rid of my waste in a recyclable state, is there a home for it in my own country or can I ship it overseas? We have never felt the need to find out how to provide a better product which might help the buyer in his or her trade.

During the 1980s and 1990s when the recycling drive was building up everyone was creating mountains of waste and wondering how to turn this waste into quick revenue. Buyers were found largely because they were taking anything and had the manpower to sort through the bales we sent them, but they could have used it better if the sellers had more understanding of their needs and costs. I wonder if even today we are giving that end of the chain sufficient thought. What has got people thinking in recent years, as we have seen, has been the introduction of stricter controls such as Green Fence which have forced people to think of China as a partner as well as a customer and to ask what a partner really needs instead of just supplying a product.

In other words we are slowly learning and therefore the success of recycling primarily depends on education and the awareness of the need for recycling. Education about the importance of recycling has been inadequate over the last 25 years. We need to increase the level of education from junior to senior levels at school, universities and then extend it into businesses and the home. There must be no excuse for failing to understand why it matters. A number of academic institutions in the Western economies are introducing these studies into the curriculum but this needs to be a global phenomenon, not one restricted to a few countries and there is an onus on the industry to take a lead here in promoting what it does.

When storms hit parts of the UK over Christmas and the New Year 2013/2014 vast areas were flooded, homes had to be abandoned and all services were disrupted or forced to shut down. One utilities company decided that while it couldn't make up for ruined Christmases it could offer something to the local people to compensate for their misfortune. Cash payments were made to the councils who used the money to pay for facilities at schools and hospitals. In order to raise the profile of waste collection and its importance some companies might consider putting something back into communities from which they profit. The point is to emphasise that the community and the recycling company are partners in the enterprise; households and businesses create waste, the waste is collected and recycled in order to produce more goods which we all need and buy only to create more waste.

This may seem self-evident but when it comes to developing countries it is not so straightforward. The profits a recycling company can readily make in the developed world are harder to come by in poor and emerging economies. Recycling only works for a private investor when the volume collected justifies the installation of a plant and the output generates revenue to meet the costs and allow a profit to be made. However in many countries in Africa where the population is not that dense per square mile – in fact there are such areas in Western economies as well – it would be expensive to collect waste. But as part of the international aid programme the richer economies should help emerging economies to build recycling operations which are appropriate to the country, its geography and its people. As we have noted elsewhere the return on trying to push from a low level of recycling compared with 'topping up' an extra percentage point on a programme which is already achieving 75 per cent is much greater. Televisions and mobiles today are a global phenomenon but are we able to recycle every phone that ends up as scrap in Africa? Are we able to recover the plastic, metal and other valuable parts from every television that is scrapped in Africa? Are we able to recycle every car that is scrapped in the deserts of Africa and the Middle East? Are we able to recover steel from all the buildings that are being demolished in the course of modernisation and new buildings coming up? Are we recovering all the metal or the plastic that we can from those sites? The answer is probably, no, because it is not cost effective. But it could be of a global benefit because that steel and plastic may not be of immediate use in the country of origin but it would be a very important raw material for many industries in Europe, America or countries in Asia.

It may be an unpopular suggestion but rather than spending so much money on combating climate change part of global international aid should be focused on waste recovery in these African countries where we can promote this industry and bring it to a sustainable level to the point where they can be equal global partners in providing usable raw material for the planet.

I say global partners because everyone from the reader of this book, collector and processor of waste to the user of raw materials is a global partner in the process of increasing the amount of waste to be recycled and decreasing the need for virgin raw material to be used instead. If a country in Africa is able to supply 100,000 scrap cars a year for recycling then they become a very important global partner in the steel supply chain and an integral partner in the recycling industry. The volumes available may not be big enough to warrant the interest of a global corporate but to help stem the drain on finite virgin material we need to spend money to acquire waste from any country in the world even when at first the immediate returns may not seem so obvious.

Investment, so often a challenge for any company seeking to start up or expand, is finding its way into the recycling industry particularly into the high profile waste-to-energy sector where equity partners and venture capitalists see quick rewards. It is not so glamorous but without collection and sorting even WTE facilities will struggle. Even if the waste is to be burned or turned into RDF the quality of furnish is just as much of a requirement if the plant is to achieve the highest possible yield. This can be considered as backward integration for the WTE sector which is likely to see the biggest slice of investment over the next ten years. There should be a more streamlined approach from collection, sorting, baling and producing the right sort of product whether for recycling or WTE.

The industry will continually be evolving and attracting investment as businesses find new and more efficient ways of reducing the waste we produce every day: tyres will be thinner but more robust and equally roadworthy, using less rubber and metal; paper perhaps surprisingly will also be thinner but with the same 'burst factor' as the paper we have been used to. It will mean that a box which can still hold 20 kg today will be made of lighter paper and when we throw it away we will be creating less waste material than we did 20 years ago. The investment may not necessarily be in the recycling process but we will have learned from the process thus fulfilling one of the three 'Rs' to *reduce* the amount of waste we create. We are not going to stop driving cars or packing our goods but at least we will be cutting down on the rubbish we discard.

This approach has an impact on other industries such as construction as we witness so much demolition and rebuilding around the world. Instead of discarding the tonnage that ends up in landfill – rubble, steel, timber and plastic – we will gradually devise more efficient ways of utilising the debris.

I would also like to raise the profile of one of the three 'Rs', *reuse*, which I feel is too often neglected. In the developing and poorer economies people reuse more than recycle for very obvious reasons but our profligate lifestyles in developed nations means we are less inclined to make use of something once it has served its primary purpose. For example once a new computer is taken out of its box the box is discarded without a moment's thought about how it might be reused. And yet it is probably made out of high density material and could easily be used for storage before it goes back into the recycling process if there were a system for collecting boxes which had been opened carefully. I don't see the reuse of boxes in the Western world as being a big business but it is the principle which I would like to promote: the idea that we take a second look at everything and ask ourselves how it can be reused before we throw something away. Everyone is familiar with the recycling logo, so maybe we should consider devising a Reuse Logo to stand alongside it to remind us.

Some items do not lend themselves to being reused because they contain a microchip and if the product goes wrong the entire board or chip needs to be replaced which is probably 80 per cent of the cost of the equipment. Factoring in the labour costs, it makes more sense to buy a new model but there are other items whose life can be extended.

A lot of pharmaceuticals have a use-by date and when those medicines are within a week of that date if they haven't been used they just get thrown away – to be on the safe side. But there are charities which collect medicines such as antibiotics and immediately send them to developing countries where they can be used safely within that last week. Those countries are constantly running short of medicines and if they can be airlifted, distributed and prescribed in time lives will be saved. The same could be applied to the mountains of food which even today get discarded, sometimes just because they are the wrong shape. But imagine the difference a crooked carrot or misshapen potato could make to the destitute. Reuse in this way becomes more important as recycling volumes increase and a growing global population demands new goods.

In my particular speciality of paper, reforestation is a primary concern but we need to do more. Instead of one new tree for one cut down we need to increase the level of planting to take into account the increased consumption of paper that we will see over the next 10–25 years. It is not sufficient to replace 10 acres of forest with 10 acres of replanting elsewhere; we have to double our efforts not least because of their direct impact on climate change.

It may seem argumentative but let me ask the question: should we be promoting the battle against climate change or should we be promoting recycling first? In recycling we are challenging the potential shortage of scarce material because we all agree that in the years ahead we will slowly be depleting the reserves of natural raw materials making them increasingly scarce. By recycling we are reducing the emission of greenhouse gases which we know has a direct effect on the climate. I would say recycling must be given a higher priority because by promoting recycling we are, for example, saving the forest and reducing carbon emissions – two important elements affecting climate change. Therefore, yes, we must talk about climate change but we should first be asking ourselves why it is happening and the answer is because we are consuming more, burning more, scrapping more. Instead, if we reuse and recycle we will have an immediate beneficial impact on the environment.

17

Seven steps

Let me gather all these thoughts together and suggest some solutions to the question we considered at the start of this short book: How are we going to clear up the waste we create? Somehow we have to develop a waste collection and recycling system which could be operated equally well in a densely populated city as well as an area where people still live far apart. Whatever we propose must be able to capture small and large volumes which may change in composition as local communities evolve and at the same time be able to offer a recyclable product which can command a good price on the open market.

The EcoHub

With the high per capita consumption and waste generation that we have in Western economies what we are failing to do is to maximise the recycling potential that we have from the waste collections in our communities. There can be multiple reasons for this including increased cost of collections, increased cost of processing, increased cost of producing the right qualities, and smaller volumes of different types of recyclable collected rendering them not worth processing. This can be seen in the plastics industry where a plant might only receive 5–10 tonnes of plastic waste – used milk containers or Coca-Cola bottles – which does not make economic sense to separate so the processor puts it all into one bale and offers it for sale as a mixed bale. But Coca-Cola bottles are PET plastic and the milk containers are HDPE, two entirely separate grades to be used for completely different end products in different manufacturing processes. In other words potentially perfectly valuable waste is being devalued right from the start.

What is the processor to do? He or she cannot invest in advanced sorting mechanisms because of the lower volumes of different kinds of material being collected. There is not enough value being generated and the same is true across any Western economies and even worse in emerging nations.

The challenge therefore is to maximise recyclables from collection streams such as these where there are three fundamental criteria for success:

1. Investment in better or more modern sorting systems that can receive and sort the multiple grades that are received as feedstock for the plant.

2. Plants with larger processing capacity of recyclable material to justify the investment made in the plants that have been created.

3. Minimum logistics costs

All these basics can only be satisfied if we decide to create within the country what I call a series of EcoHubs. I have no doubt that in time these will be seen as the obvious method of collection and recycling, no different from the principle of having one multi-storey car park in a town where everyone is trying to do the same thing – park their cars.

The EcoHub should be created to cater for a waste collection zone of say one to two million people which would generate enough material to support the hub. These hubs would be served by satellite stations which would collect, receive and feed directly into the EcoHubs where there would be giant plants capable of sorting the collections into better refined grades and of acceptable international quality. Assuming one million people generate a quarter of a million tonnes of waste per annum, a city the size of Greater London might need six EcoHubs. Each hub would have the most advanced systems for handling all waste including sorting plastic, sorting and shredding metal and handling paper, glass and timber waste.

Recycling as an industry only succeeds when transport, volume and technology are combined in a most efficient manner and only an EcoHub can combine all three of these. In the paper industry you might have MRFs of different classes, some high quality, some medium and some low which as a result are producing different grades of material and not all will find a ready buyer. But if we have one large processing centre – the EcoHub – and everyone filters their material through the system you might have various plants capable of handling quarter of million tonnes of paper, 50,000 tonnes of plastic, 20,000– 30,000 tonnes of metal a year, glass and organic waste. Not only can it all be handled in a more efficient manner using the best equipment but even small individual quantities can be received into the facility; they would simply be sorted and treated with larger waste collections from a wider catchment area.

At the moment sorting stations find that they cannot recycle all the material they have received and because the nearest specialist plant might be so far away they choose to landfill the waste rather than recycle it thereby rendering the whole collection purpose both pointless and wasteful.

EcoHubs would be substantial enterprises in state-of-the-art buildings all operated behind closed doors containing odours as well as windblown waste.

They would therefore require substantial investment by either private sector operators or in a public/private enterprise. But as they would be regionally based serving a large community, the investment would be compensated by the increased volume they received. The current process I fear may ultimately be unsustainable; where even medium- and large-sized companies have made investments in modern MRFs but are not able to sustain the operations of those facilities because they are not able to get the right mix of the recycling material that they need to run them. By contrast the EcoHubs will have enough feedstock supplied by small, medium and large collection stations gathering in material which they know will always be accepted by their nearest EcoHub. The hub becomes one big umbrella where multiple processing is taking place – a one-stop shop for all recyclables.

Technology

Over the past 25 years there has been a never-ending development of new technology in the recycling industry. We have moved from rudimentary collections and dumping to computerised, laser-controlled sorting systems and highly mechanised treatment and biological production. There can and will be no let-up, not least because such sectors as RDF and WTE continue to dangle the carrot of financial rewards. I always advocate focusing on the beginning of the chain: the waste collection. Vast savings can be made in one simple area which is always overlooked: the moisture content in our waste. When we transport drinks cartons, aluminium cans, milk cartons or even paper left for collection outside overnight we are also transporting unwanted liquid. I foresee that there will soon be some kind of heating system which will extract excess moisture perhaps turning it into steam which can be harnessed as energy to run the plant instead of taking electricity from the grid with obvious eco-friendly benefits.

If we can reduce the moisture in our cargoes we will make substantial savings in logistics costs, reduce the carbon footprint of the journey and we will cut losses at processing plants. The moisture in the recyclable raw material represents a total book loss in the course of recycling for the manufacturer who is buying paper or aluminium, not water. The figures may surprise people. A cardboard container may have as much as 25 per cent moisture content, so for every 1,000 tonnes of cardboard sold, the water content could be around 250 tonnes. There are restrictions; officially the global specification allows for 10–12 per cent moisture, but it is difficult to monitor. The same thing applies with aluminium cans where moisture content could be up to 4 per cent, which is much higher than the level at which manufacturers will buy those cans. One can pay well over $1,000 a tonne for aluminium cans, so 2–3 per cent extra weight from the moisture content equates to $20 or $30 per tonne just for water which is of no benefit to anybody.

The biggest advantage from a technological point of view with the EcoHub is that it provides multiple sorting, collecting centres of various recyclables within the hub all of which can be constantly updated – plastics sorting would be just one element of the whole – so the cost of upgrading to accommodate new technology or even a change in international demand can be spread across the EcoHub enterprise.

Logistics

One of the biggest costs in the recycling industry is logistics and transportation. In international movement part of this cost is borne by the shipping lines who are constantly working on refined models to reposition their containers to be as near as possible to the export loading point. For us in the recycling industry at times the transportation cost paid for moving recyclables from the point of arising to the point of reprocessing is sometimes the same as the value of the material itself or even in some instances such as paper, wood or glass, actually more than the value of the material. Therefore it becomes a big challenge for the recycler to find ways of reducing the cost of transportation.

Alongside transportation there is the expense of storage as recyclables sometimes need to be stored because of low volumes. But in an EcoHub environment there is no need to store because there will always be a plant nearby that can swallow every kilo of the recyclables that is delivered.

Despite the constant search for ways to move the material as cheaply and as efficiently as possible, a truck might typically travel many miles in the course of the day just to collect 10 tonnes of material; this is neither cheap nor efficient as it not only increases the cost but also the effective carbon footprint of the journey. The challenge for us is to find ways of reducing the miles travelled for every tonne of recyclables we collect. The EcoHub will help us move in that direction. It may be necessary to travel further to collect the material but there will be less travel to reprocess it in order to get a higher yield and a better quality product as the EcoHubs will be strategically located across the country.

As the bulk of waste is transported across the oceans, shipping lines are constantly seeking new ways to position their containers in the most cost-efficient way because they are not able simply to leave them close to the loading site. They have to collect the loaded container and take it back to the port. EcoHubs will help reduce international transportation costs because the containers will be positioned close to the EcoHubs. If there are 20 hubs there are 20 known and reliable destinations for the containers which not only cuts the repositioning costs of the shipping lines but it also cuts the cost of loading and freight rates.

I would even advocate positioning an EcoHub facility at the ports where recyclable materials can be processed in a transport friendly manner, whether loading of metals in bulk or even the storage of material awaiting shipment because the low value of some of these items means it is not always economical to store them in existing facilities near the ports.

I would go a step further and suggest that there should be special terminals to handle recyclable materials at a preferential rate at ports. On the whole this tends to be low-cost material possibly even worth less than the sea freight as we have noted. In other words in such exceptional cases logistics becomes the primary business not the material that is being shipped. A dedicated recycling hub at or adjacent to the main ports around the world would seem logical and this will become even more important in the years ahead. The increased size of the ships should mean a reduced unit cost of shipping a tonne of material. While prices have certainly come down from the sea freight levels we saw in 2005–2007 the indications are that we will probably be seeing a marked rise in sea freight costs in the future.

Value

European Union legislators are gradually recognising recyclables as a raw material; unfortunately so far they have not recognised all the waste that is produced as a raw material. For example while some non-ferrous items have been accepted as 'End of Waste' (i.e. a raw material), waste paper has still not been accepted and yet waste paper is the largest volume of waste that is processed the world over. There is no doubt in my mind that the EU should recognise waste paper as achieving raw material status once it has gone through the processing chain and a properly defined quality of waste has been produced. Failure to recognise the value of this and other waste categories is a handicap in the global trading of waste and achieving the global loop I have spoken about and which is so vital.

The conditions being imposed are too restrictive and there needs to be more understanding of how, for example, the paper recycling industry operates and how difficult it is to achieve 98 per cent purity. Parliamentarians are debating controlling contamination to very low levels but I would argue that we are dealing with collections that are waste treated and sorted within agreeable tolerances which the industry is prepared to accept and if we can come within the acceptable limits it should be treated as a raw material. Not just any raw material but an essential raw material.

Indeed all waste should be recognised as an essential raw material with a value without which the world would probably struggle to function effectively as everything we use on a day-to-day basis depends to a greater or lesser extent on the recycling industry.

Harmonise

This is a self-evident but strangely impossible step for the recycling industry to take exacerbated by our governments and at a local level our cities and local authorities. We refuse to sing from the same hymn sheet while professing to support the concept of recycling. At government level we impose different standards and different requirements. Inspection even pre-inspection regulations seem to run counter to the desire to maximise our waste, and the three 'Rs' of Reduce, Reuse and Recycle are forgotten. While different countries have different requirements every kilo of waste should find a home even if it is to end up in a WTE furnace. There must surely be common ground in recognising that it is important for this industry to grow, increase the use of recyclable material and decrease the usage of natural raw materials. We must promote recycling and this can only be done if we promote movement of recyclables worldwide and that movement will only take place when we have common, well-defined legislation which is globally harmonised and we have inspection regimes which are properly implementing internationally accepted codes of practice in terms of quality and grade descriptions.

We could add recycling systems upstream and downstream. Downstream we must focus on how waste should be collected from our doorsteps, from supermarkets and other businesses. Should it be a one bin collection or multi bin? There must be a right and wrong way to something as straightforward as collecting rubbish. Should it be once a week or once a fortnight? We have discussed moisture content in paper and the fact that the longer it is left outside the more moisture it is absorbing, which will have a direct effect on the yield the fibre will generate. Therefore what is the balance – once a week, twice a week, one bin or two or more? I have a question for our political masters – how can it make sense for two authorities operating next door to each other to operate different systems? One is surely right and one is wrong. The reasons these things happen of course are all to do with the collection and processing systems which the local contractors operate in that community. Maybe the MRF if there is one is not sophisticated enough to handle multi-bin collections or cannot sort commingled waste streams.

We must ask ourselves difficult questions: should the game be harmonised at a regional or national level, even global level rather than at a small community level? What impact will it have on the cost of collection or even on the existing contractors? Another question if we continue to ignore these anomalies is: what impact will it have on the quality of the product our region is producing and will there be a market for it? Green Fence controls and others like it suggest standards are going to rise not fall so are we prepared or will what we collect simply end up in landfill which defeats the entire purpose of collecting in the first place.

Prioritise

The world is changing, more of us are moving into the cities, so what will be the impact of this increased urbanisation on our waste stream? Governments must evaluate their priorities and the answers will not be the same everywhere. There will be vast cities imposing huge burdens on infrastructure and they will be producing more waste than we can even imagine today – there will have to be a coordinated system.

At the same time there will be populations spread thinly through the countryside, deserts and vast tracts of open land. Not everyone will find work or even a home in the urban sprawls, so will we just have to ignore them?

The reason I suggest *prioritising* as a key step is that there is no thought, let alone coordinated thought into how we will tackle waste in the future. The problem is going to be so big that if not an EcoHub then some centralised, systematic collection and processing system will have to be devised. The speed with which New York, London, Dubai, Cape Town and Mumbai are growing means that we do not have the luxury of opting for *à la carte* solutions: one for my district, one for yours.

Global economies should start promoting and challenging the way their waste streams are being processed. Are the arisings going directly to landfill? If so, how can this be reduced? All economies should be looking at ways to downsize the volume going into landfill, recapture it and provide incentives to individuals and to businesses to embrace and promote increased recycling.

Education

Lastly, but surely first and always, we must educate. Downstream from the creator of waste and upstream to the collector and waste processor and final reprocessor of the final product everyone is part of the story. Even in our daily lives we are involved.

There needs to be education in terms of understanding what should be recycled, what is a recyclable, how the process works, what does recycling generate as an end product, and how we use it in our lives. Further upstream we need to ensure that the recycling company which is processing the waste must understand that they should only be producing grades and quality of material which the ultimate user or reprocessor can use in the most cost-effective manner so their yield is increased. The recycling plant must have a detailed understanding of the requirements to produce a recyclable raw material and why shipping contaminated raw material half-way round the world is a fool's game. In short recycling becomes a choice that we all support.

What does the future hold for our industry? In the upper tier of the metal business because they are high value items there will be an ongoing demand as infrastructure is developing worldwide. So the need for steel and non-ferrous metals will be quite dominant over the next 25 years but we will have peaks and troughs of supply and demand for a short period as we continue through austerity drives and construction slow down which will see steel prices fluctuate. But fundamentally demand will remain strong and those industries in America and Europe will continue to grow at a steady pace.

As for the middle tier of paper and plastics we are seeing very clearly a two-pronged approach: we will only recycle those products if they are making money for us, if they are not we will use RDF or WTE options.

But we must not lose sight of global trade and the importance of paper within the recycling industry: a) it is the highest volume of waste that is recycled and traded round the world; b) if we do not recycle it and burn it instead we have lost it for good; and c) we may certainly have surplus recyclables in America but there will be a demand elsewhere and we must allow those markets to thrive and become members of our supply chain in terms of finished goods. More importantly recycling generates significant employment across the world by the time it is picked up at the doorstep until the time it is processed at the other end – collection, sorting, baling, logistics which includes the domestic and international movement of waste to mills. There is a great deal of investment in shipping; the recycling industry makes up an important slice of the core cargo – or base cargo as they call it – albeit low revenue but it allows the shipping line to make use of what would otherwise be empty containers returning to pick up more goods.

These aspects must not be ignored. Focus needs to be on how technology can be better developed to deal with waste so we are able to produce a user-friendly product in a cost-efficient manner.

Reinhold Schmidt, President of the BIR Paper Division, identifies the issue at the heart of the recycling debate:

> The argument that 'it is much too expensive!' will in the future be irrelevant. What is expensive, if we have no air to breathe and no more water to drink, because water is source of all life? If we really want to calculate everything, we come to the conclusion today that it is better to reuse all substances again… We all have to think of the generations that will inhabit this planet after us. We no longer have a lot of time.

At the end of the chain is the buyer and the buyer demands quality which will always be expensive to achieve. But quality is the watchword which will set the trend for the next 25 years and we will depend on recycling to produce quality goods. That is our future.

Recycling facts and figures

Upside:

Ferrous metals
- Almost 40 per cent of the world's steel production is made from scrap.
- Recycling one tonne of steel saves 1,100 kg of iron ore, 630 kg of coal, and 55 kg of limestone.
- CO_2 emissions are reduced by 58 per cent through the use of ferrous scrap.
- Recycling one tonne of steel saves 642 kWh of energy, 1.8 barrels (287 litres) of oil, 10.9 million Btus of energy and 2.3 cubic metres of landfill space.
- Recycling steel uses 75 per cent less energy compared with creating steel from raw materials – enough to power 18 million homes.
- Steel recycling uses 74 per cent less energy, 90 per cent less virgin materials and 40 per cent less water; it also produces 76 per cent fewer water pollutants, 86 per cent fewer air pollutants and 97 per cent less mining waste.
- Steel automobile frames contain at least 25 per cent recycled steel and a typical electrical appliance will usually be made of 75 per cent recycled steel. Steel cans consist of at least 25 per cent recycled steel.

Non-ferrous metals
- Almost 40 per cent of the world's demand for copper is met using recycled material.
- At present, approximately 30 per cent of global zinc production comes from secondary zinc.
- Over 80 per cent of the zinc available for recycling is eventually recycled.

Aluminium
- Of an estimated total of 700 million tonnes of aluminium produced since commercial manufacturing began in the 1880s, about 75 per cent of this is still being used as secondary raw material today.
- One tonne of recycled aluminium saves up to 8 tonnes of bauxite, 14,000 kWh of energy, 40 barrels (6,300 litres) of oil, 238 million Btus of energy and 7.6 cubic metres of landfill.
- The energy saved by recycling one tonne of aluminium is more than enough to power a US household for a whole year (the average US household uses about 10,000 kWh per year).
- Recycling aluminium uses 95 per cent less energy than producing aluminium using raw materials.

- Recycling one aluminium can saves enough energy to power a 100-watt bulb for almost four hours.
- A used aluminium can is recycled and back on the grocery shelf in as little as 60 days.
- For every single can manufactured using virgin ore, the same amount of energy used will produce 20 recycled cans.
- The aluminium drink can is the world's most recycled container – more than 63 per cent of all cans are recycled worldwide.

Copper

- Copper's recycling value is so high that premium-grade scrap holds at least 95 per cent of the value of the primary metal from newly mined ore.
- Recycling copper saves up to 85 per cent of the energy used in primary production.
- In order to extract copper from copper ore, the energy required is approximately 95 million Btu/tonne. Recycling copper uses much less energy, about ten million Btu/tonne.
- By using copper scrap, we reduce CO_2 emissions by 65 per cent.

Zinc

- The average car contains up to 10 kg of zinc in its galvanised body panels. When they are discarded, these panels can be readily made into new parts of identical quality.
- Total recovery of zinc within the non-ferrous metals industry amounts to 2.9 million tonnes, of which 1.5 million are new scrap or process residues and 1.4 million are old scrap.
- Secondary zinc production uses 76 per cent less energy than primary.
- Nearly 70 per cent of zinc from end-of-life products is recycled. Old zinc scrap consists primarily of die cast parts, brass objects, end-of-life vehicles, household appliances, old air-conditioning ducts, obsolete highway barriers, and street lighting.

Lead

- 50 per cent of the lead produced and used each year throughout the world has been used before in other products.
- Today, about 80 per cent of lead is used in acid batteries, all of which is recoverable and recyclable. Some countries boast a 100 per cent recycling rate and most are capable of the same result.
- Using secondary lead instead of ore reduces CO_2 emissions by 99 per cent

Tin

- Global tin production amounts to 350,000 tonnes of which 50,000 tonnes is produced from scrap and other secondary sources.
- Primary production of tin requires 99 per cent more energy than secondary production

Paper

- Recycling one tonne of paper saves 24–31 trees, 4,000 kWh of energy, 1.7 barrels (270 litres) of oil, 10.2 million Btus of energy, 26,000 litres of water and 3.5 cubic metres of landfill space.

- Burning that same tonne of paper would generate about 750 kg of carbon dioxide.
- Recycling paper saves 65 per cent of the energy needed to make new paper and also reduces water pollution by 35 per cent and air pollution by 74 per cent.
- Recycling one tonne of corrugated containers saves 390 kWh of energy, 1.1 barrels (176 litres) of oil, 6.6 million Btus of energy, and 5 cubic metres of landfill.
- Recycling cardboard requires only 75 per cent of the energy required to make new cardboard.
- Europe recycled 71.7 per cent of its paper in 2012.[82]

Textiles

- Of all collected textiles, approximately 50 per cent are reused and 50 per cent are recycled.
- If everyone in the UK (60 million people) bought one reclaimed woollen garment each year, it would save an average of 1,686 million litres of water and 480 tonnes of chemical dyestuffs.
- Nearly half of discarded textiles are donated to charities. About 61 per cent of clothes recovered for second-hand use are exported.
- In many African countries, over 80 per cent of the population dress themselves in second-hand clothing.
- With the reuse of recovered materials in manufacturing processes or in consumption cycles, there is a strong decrease of CO_2 emissions compared with the production of virgin materials. Here is an example of the environmental benefits derived from a study by the University of Copenhagen (research sources 2008) which shows the environmental advantages resulting from the collection of used clothing. By collecting 1 kg of used clothing, one can reduce:
  - - 3.6 kg of CO_2 emissions
  - - 6,000 l of water consumption
  - - 0.3 kg of the use of fertilisers
  - - 0.2 kg of the use of pesticides

Stainless steel

- Recycling one tonne of steel saves 1,100 kg of iron ore, 630 kg of coal, and 55 kg of limestone.
- An average stainless steel object is composed of about 60 per cent recycled material.
- Approximately 90 per cent of end-of-life stainless steel is collected and recycled into new products.

Plastics

- One tonne of recycled plastic saves 5,774 kWh of energy, 16.3 barrels

---

82   European Recovered Paper Council (ERPC)

(2,604 litres) of oil, 98 million Btus of energy, and 22 cubic metres of landfill.

- There is an 80 to 90 per cent reduction in energy consumption by producing recycled plastic compared with producing plastic from virgin materials (oil and gas).
- Recycling a single plastic bottle can conserve enough energy to light a 60-watt bulb for up to six hours.
- Recycling five PET plastic bottles produces enough fibre for one T-shirt.
- Recycling 100 million cell phones saves enough energy to power more than 194,000 US households for one year.
- Worldwide trade of recyclable plastics is valued at $5 billion per year and is estimated to represent a total of 12 million tonnes.
- Europe recycled 21.3 per cent of plastic waste during 2008 representing about 5.3 million tonnes.
- A recent study shows that if all landfilled plastics waste are recycled or re-covered into energy, then 7 per cent of EU quota of carbon emissions will be fulfilled.

Tyres

- The oil required to retread a tyre is 20 litres less than the oil needed to manufacture a new tyre. With commercial vehicle tyres, the savings are even greater, estimated to be about 68 litres per tyre.
- Retreading a tyre costs anywhere from 30 per cent to 70 per cent less than manufacturing a new tyre.
- Over 90 per cent of all aircraft tyres are retreads.
- Scrap tyres used as fuel can produce the same amount of energy as oil and 25 per cent more than coal.

Downside:[83]

- 50 per cent of all camels and cows in India and the United Arab Emirates (UAE) die from eating plastic waste.
- With a population of around 29 million, Saudi Arabia generates more than 15 million tonnes of solid waste per year. The per capita waste generation is estimated at 1.5 to 1.8 kg per person per day.
- The potential annual sewage sludge and septic tank wastewater generated in Amman can be estimated at 85,000 tonnes of dry matter. Jordan also generates a significant amount of animal manure because of its large animal population in the form of cattle, sheep, camels, horses, etc.
- It is estimated that 100 million tonnes of waste is floating in the Pacific Ocean covering an area twice the size of the continental United States. This 'plastic soup' off the California Coast across the northern Pacific, past Hawaii and almost as far as Japan is held in place by underwater currents.

---

83   Other Sources include: Bureau of International Recycling, Greenwaste.com, 60 Minutes, UK Government, US Government (EPA), Scientists for Human Rights, RandomHistory.com, EU EuroStat, CNBC, Plasticoceans.net

- Fourteen billion lbs of garbage, mostly plastic, is dumped into the ocean every year.[84]
- From 1946 through 1993, 13 countries (14, if the USSR and Russia are considered separately) used ocean disposal or ocean dumping as a method to dispose of nuclear/radioactive waste.
- In 2010, the total generation of waste from economic activities and households in the EU-27 (i.e. excluding Croatia which joined the EU in July 2013) amounted to 2,502 million tonnes.
- Among the waste generated in the EU-27 in 2010, some 101.3 million tonnes (4.0 per cent of the total) were classified as hazardous waste.
- The US produces a quarter of the world's waste despite the fact that its population of 300 million is less than 5 per cent of the world's population, according to 2005 estimates.[85]
- The United Nations and other agencies estimate worldwide annual waste production at more than one billion tonnes, and some estimates go as high as 1.3 billion.[86]
- Global MSW generation levels are expected to increase to approximately 2.2 billion tonnes per year by 2025.
- OECD countries produce almost half of the world's waste, while Africa and South Asia regions produce the least waste.
- About one third of all food produced for human consumption goes to waste.[87]
- In 2011, the world threw away 41.5 million tonnes of electrical equipment, and this is expected to rise to 93.5 million tonnes by 2016.
- Approximately 130,000 computers are thrown out every day in the US. Over 100 million cell phones are thrown out annually.
- It was estimated in 2005 that 75 per cent of electrical and electronic goods imported into Lagos were junk.
- South Africa produces nearly 67 million cubic metres of waste per year.
- South Africa causes 98 per cent of air pollution in Africa and 86 per cent of waste in Africa – mainly due to the heavy impacts of mining[88]
- Over 80 per cent of items buried in landfills could be recycled instead.[89]
- Americans throw out around 40 percent of their food worth about $165 billion a year according to the National Resources Defense Council.
- In 2011, the US generated 2.2 million tonnes of e-waste, according to a US EPA-funded report, making it the fastest-growing segment of the solid waste stream.

84 Orme, Helen (2008) *Earth in Danger: Pollution.* New York, NY: Bearport Publishing
85 *Forbes*
86 *Forbes*
87 United Nations Food Agency
88 Scientists for Human Rights
89 Jakab, Cheryl (2007) *Global Issues: Clean Air and Water.* North Mankato, MN: Smart Apple Media

- In 2008 around one billion End-of-Life Tyres (ELTs) were being produced globally each year. A further four billion were estimated to be held in stockpiles and landfills, according to a report by the World Business Council for Sustainable Development. Around 1.5 billion new tyres are produced annually.

- The average office employee throws away 360 lbs of recyclable paper each year.[90]

- In 2008, a total of 289 million tonnes of waste was generated in the UK from various sources including construction (35 per cent), mining (30 per cent), industrial and commercial (23 per cent), household (11 per cent) and secondary sewage waste (1 per cent)

- UK Department for Environment, Food and Rural Affairs (Defra) named and shamed Ashford as the worst recycler in the country in 2012, for the third year in a row. Ashford recycled 14 per cent of household waste, compared with a national average of 43 per cent.

- Households in Europe and the US alone get through 40 billion cans of food a year, according to the Can Manufacturers Institute in Washington, DC.

- Over one million seabirds are killed by plastic waste per year. Over 100,000 sea mammals and countless fish are killed per year due to pollution.[91]

- A plastic bag has an average 'working life' of 15 minutes.

- Americans throw away 25,000,000 plastic beverage bottles every hour!

- The average person generates 4.5 lbs of trash every day – about 1.5 tonnes of solid waste per year. Although the EPA estimates that 75 per cent of solid waste is recyclable, only about 30 per cent is actually recycled.

- In 2004, 55 billion aluminium cans were landfilled, littered or incinerated, that's nine billion more than were wasted in 2000. This is enough cans to fill the Empire State Building 20 times. It is also a quantity equivalent to the annual production of three to four major primary aluminium smelters.

90   Wehr, Kevin (2011) *Green Culture: An A-to-Z Guide.* Thousand Oaks, CA: Sage
91   Gifford, Clive (2006) *Planet under Pressure: Pollution.* North Mankato, MN: Heinemann-Raintree Library

Index

THE BUREAU OF INTERNATIONAL RECYCLING

The founding fathers of BIR may have been European but their vision was global from the very outset. The organisation was founded shortly after the Second World War on the principles of rebuilding the bridges sundered by global conflict and of free and fair trade without exception or exclusion. BIR has held fast to these precepts ever since, defending the right of recyclers around the world to join in trade with each other so that secondary raw materials find their way to where they are most needed.

Nowadays, companies outside the EU make up approaching half of our membership: Turkey and the Middle East, for example, provide us with upwards of 10 per cent of our members while a further 20 per cent are based in China, India and other parts of Asia. And it is the non-EU, non-North American proportion of our membership that has been growing most rapidly over recent years.

Even during the last decade, a period which has encompassed one of the steepest economic downturns of the modern era, we have seen our membership numbers soar by 50 per cent. In effect, BIR has never been so well supported and so global in its coverage. We now have around 900 member companies as well as more than 40 national member associations, in total spanning well over 70 countries. It was the first, and remains the only, recycling organisation that can rightly lay claim to the adjective 'global', providing the framework to promote the development of recycling in emerging as well as in developed markets.

Increasingly, the recycling industry is a global community sharing many of the same ideals, aspirations and challenges. Through regular interaction within BIR, a world of expertise can be brought to bear not only on recycling problems but also opportunities. BIR looks to foster a spirit of inclusiveness to the benefit of recycling globally.

With the collaboration of the generations of leaders, BIR has built – over many years – the flexibility and pedigree to make an impact at both the macro and micro levels. In addition to promoting free and fair trade, we use our time with these leading decision-makers to push the need for environmentally sound management of resources, greater use of recycled materials worldwide and design of products with recycling in mind.

BIR's face is recognised in all of the corridors of power as a respected representative of the recycling industry with whom it is possible to engage in productive debate. BIR is, at once, the single mouthpiece for the global recycling industry as well as its staunchest advocate.

Our flagship Conventions are one of the most overt benefits of BIR membership in that they provide members with what is universally acknowledged to be one of the highest-quality global networking opportunities, perfect for making important new contacts and for conducting business. At the same time, these events also attract expert speakers from around the world and across the recycling spectrum, providing attendees with updates on global market developments, no matter whether their trading interests lie in ferrous, non-ferrous or stainless scrap, electronic scrap, recovered paper, plastics scrap, used textiles or tyres.

But membership of BIR means so much more than a conference pass. It is a badge of honour that can open doors in dealings with traders, consumers and officialdom around the world. Through the BIR Code of Conduct, we require our members to meet exacting standards of commercial behaviour so that business partners can feel reassured that they are dealing with a company of substance which holds itself to the highest business practices and ethics.

And in a further bid to smooth trade between the many different companies and cultures that come under the BIR umbrella, we have recently revised our Arbitration and Conciliation service so that disputes involving members anywhere in the world can be resolved more quickly, consistently and cost-effectively by having the case heard by independent experts from within the recycling industry itself.

The knowledge and expertise shared at our Conventions are only part of BIR's role to inform. Over the last six years, we have been gradually increasing our data-gathering efforts so that we can argue the recycling industry's case from a position of greater statistical certainty. For example, a BIR-commissioned report broadcast to the world the compelling fact that the existence of our industry prevents the emission of a greater quantity of greenhouse gases than are emitted by the airline industry worldwide. Such nuggets of information are helping to make the world's decision-makers sit up and take greater notice of the recycling industry.

Some of the information compiled and published by BIR is designed to promote industry-leading standards so as to increase the environmental soundness and efficiency of operating recycling plants. For example, the BIR has assisted members and other recyclers with three important publications: 'Tools for Environmentally Sound Management', 'Tools for Quality Management' and 'Tools for Occupational Health & Safety Management'. These have been formulated so that recyclers may understand, implement and then demonstrate the management system that suits their needs or that they are required to have to be in business. BIR makes these documents available to all recyclers via the internet.

Another hot topic on which BIR is actively engaged concerns the rising tide of scrap metals theft and fraud. BIR has adopted a proactive approach by using the services of the International Maritime Bureau, or IMB, a crime-fighting specialist within the International Chamber of Commerce. BIR members provide the IMB with feedback on thefts or suspected frauds, the details of which are duly checked out by the IMB for the purposes of its own database-building. The IMB is then in a position to send alerts to BIR members relating to, for example, specific ports or companies in the logistics chain. IMB can provide a sanitised report of any information it holds on a company, which may help to provide reassurance over an unknown company by showing its past trading activities.

BIR stands ready when called upon to work shoulder to shoulder with like-minded and similarly motivated recycling companies and industry representatives – anywhere in the world – in the pursuit of our mutual goals.

Alexandre Delacoux
BIR Director General

---

1   BIR Study on the Environmental Benefits of Recycling, 20